Doctor Dolittle's Delusion

Doctor Dolittle's Delusion

Animals
and the Uniqueness
of Human Language

Stephen R. Anderson

With illustrations by Amanda Patrick

Yale University Press / New Haven & London

Designed by Mary Valencia.
Set in Cochin type by Tseng Information Systems, Inc.
Printed in the United States of America.

Library of Congress Cataloging-in-Publication Data
Anderson, Stephen R.
Doctor Dolittle's delusion : animals and the uniqueness of human language.
p.   cm.
Includes bibliographical references (p.).
ISBN 0-300-10339-5 (c : alk. paper)
1. Animal communication.   2. Language and languages.   I. Title.
QL776.A5199   2004
591.59—dc22
2004044309

A catalogue record for this book is available from the British Library.

The paper in this book meets the guidelines for permanence and durability
of the Committee on Production Guidelines for Book Longevity
of the Council on Library Resources.

10   9   8   7   6   5   4   3   2   1

for the Bunnies

Already I see gray hairs showing at your temples, Tommy. If you try to write down everything the Doctor did, you'll be nearly my age before you've finished. Of course, you're not writing this book for the scientists exactly; though I often think since you are the only person so far—besides the Doctor—to talk animal languages at all well, that you ought to write something sort of—er—highbrow in natural history. Usefully highbrow, I mean, of course. But that can be done later, perhaps.

—Polynesia the parrot, from *Doctor Dolittle's Zoo*

⇥ Contents ⇤

↠ Preface ↞

For invaluable information about animals and their ways of communicating, I am grateful to colleagues and students in numerous classes devoted to this material. Years ago Greg Ball, Moise Goldstein, Stewart Hulse, and I participated in a seminar at Johns Hopkins University that made me want to explore these matters and their relation to human language in greater detail.

That interest became much more focused when Letitia Naigles and I offered a course on communication and language abilities in animals at Yale University in 1996. I have benefited greatly from Letty's contributions to that class, which she will see reflected here in many places (especially in Chapter 10). Students in subsequent incarnations of that course have helped me with their insightful questions (as well as such contributions as Figure 1.1). I am also indebted to Alison Richard (for showing me how interesting lemurs are) and David Watts (for much information about monkeys, chimpanzees, and gorillas—too little of which shows up in these pages).

Nearly all of this book is drawn from the existing literature, rather than from investigations I have conducted myself. While I have tried to acknowledge in the notes particularly obvious sources on which I have drawn, a full documentation of everything here would be far too distracting for the non-specialist reader. I must make special mention of the work of Marc Hauser, however, particularly his comprehensive survey *The Evolution of Communication*. Many of my unattributed and otherwise offhand remarks about animals draw on this work, and any potentially serious student of the subject should certainly have a copy.

My goal in this book is not comprehensive coverage of everything that is known about animal communication. Rather, I discuss systems and species for which a comparison with human language seems productive. The reader with a broader interest in animal communication can consult such general works as Bradbury and Vehrencamp's *Principles of Animal Communication.* In-depth information about birds will be found in Kroodsma and Miller's *Ecology and Evolution of Acoustic Communication in Birds;* Gerhardt and Huber provide a current survey of research on *Acoustic Communication in Insects and Anurans.* None of these works provide much information about *human* communication, but each gives much more detail on the species that are its subjects than I could possibly cite.

I am concerned here with a specific cognitive ability in animals—communication—and I cannot do justice to the vaster literature dealing with animals' cognitive capacities in general. While it has often been fashionable to maintain that animals are purely mechanical systems with no mental lives to speak of, this view has little if anything to recommend it in light of the masses of evidence accumulated by ethologists and other students of the lives of animals. The work of Donald Griffin has been particularly influential in this regard. His book *Animal Minds* is an excellent place to explore some of the issues I cannot go into here, as is Hauser's *Wild Minds.*

This book has benefited enormously from comments by Janine Anderson-Bays, Diane Brentari, Marc Hauser, Norbert Hornstein, Paul Moore, Duane M. Rumbaugh, and three anonymous reviewers. Louis Goldstein provided some of the sound files used for analysis in Chapter 5, and Susan Fischer passed along valuable evidence to which I refer in Chapter 9.

I am grateful to Amanda Patrick for the charming drawings with which some of the more tedious stretches below are enlivened. Finally, I express my appreciation to my editors at Yale University Press. Apart from immensely helpful suggestions, including ways to leaven and generally improve my rather rigid prose, Jean Thomson Black contributed greatly by seeing this book through a number of unanticipated difficulties—and by having faith in it. It certainly would not have come into existence without her aid and advocacy. Vivian Wheeler made vast improvements in the style, for which the reader will be as grateful as I am.

1

Animals, Language, and Linguistics

"Why don't some of the animals go and see the other doctors?" I asked.

"Oh Good Gracious!" exclaimed the parrot, tossing her head scornfully. "Why, there aren't any other animal doctors—not real doctors. Oh of course there *are* those vet persons, to be sure. But bless you, they're no good. You see, they don't understand the animals' language; so how can you expect them to be of any use? Imagine yourself, or your father, going to see a doctor who could not understand a word you say—nor even tell you in your own language what you must do to get well! Poof!—those vets! They're that stupid, you've no idea!"

—*The Voyages of Doctor Dolittle*

Hugh Lofting's fictional Doctor Dolittle certainly was kindly and well-meaning—indeed a great man, and one who accomplished much for the animals he loved. Nonetheless, he must have been suffering from a serious misconception: the delusion of this book's title. Merely believing that all animals have ways of communicating with one another would have been an eminently sensible position for the renowned naturalist to take. Where he (together with his friends in the books—and all too many others, down to the present day) went off the track was in equating these abilities with the human faculty we call *language.* In pointing this out, I certainly do not mean to denigrate the good Doctor and his colleagues, but as I am sure he would have acknowledged, scientific truth cannot be ignored.

For there is indeed a science that can sensibly establish the fact of the matter: *linguistics,* a field whose relation to language and languages is every bit as principled as the relation of, say, geology to rocks, minerals, and mountains. Over the past century or so, a scientific understanding of human natural language has developed. It is specialized and technical in its relation to its subject matter, with methods and results that are not instantly apparent but are nonetheless well supported by a long tradition of inquiry. People sometimes are incredulous to hear linguists suggest that what they are doing is somehow comparable to physics, but a great deal that is known about language has a genuinely scientific character, and can be appreciated only on the basis of an understanding of the relevant science.

Every normal human being raised under normal conditions has fluent control of at least one language. It is tempting to conclude therefore that the organizing principles of language should be evident to anyone who chooses to think about them. But this is a mistake, and one that seriously underestimates the complexity of the matter. Hardly anyone would argue that golfers or baseball players, adept as they are at controlling and predicting the flight of balls, must as a consequence know everything there is to know about the physics of small round objects. The systematic study of language similarly reveals properties that are far from self-evident.

When examined scientifically, human language is quite different in fundamental ways from the communication systems of other animals. Still, there are interesting and sometimes quite detailed similarities and we can learn important things about the one by studying the other. In the end, though, the differences are so important that we must not obscure them. What other animals do is not just their own variant of our human talk, in the way Japanese is a variant of what English is. Pursuit of that analogy makes it impossible to understand the basic nature of human language or to see animal communication systems in their fascinating richness rather than as some pale imitation of English.

Indeed, the central question of this book might be: To what extent is our use of natural language a uniquely human ability? In answering I want to convey some of what the modern science of linguistics teaches us about the basic properties of language. To put the result of that inquiry into some sort of perspective, I take other communication systems seriously as well in presenting what is known about their basic properties. I explore two fascinatingly rich and detailed areas of inquiry: animal communication and cognition on the one hand and human natural language on the other. Although they differ in fundamental respects, we can learn a great deal by comparing them.

For much of human history, use of language has been cited as a characteristic that defines human beings and sets us apart from all other animals. Since the 1970s, though, the purported uniqueness of this capacity has come under attack. It seems fair to say that the current understanding in the popular press is that the conception of language as an ability limited to humans is not only outmoded but even a kind of prejudice that science has shown to be wrong—along with many other supposed differences between humans and nonhumans such as the use of tools and the cultural transmission of knowledge and behavior. Other animals, this opinion holds (specifically various higher apes, such as chimpanzees), can be taught a human language and can use it to communicate. And anyone who says otherwise is a rank species-ist.

Consider a review article that appeared in the *New York Times Book Review* not so very many years ago. Its thrust is that we humans ought to be kinder to our ape cousins, and I have no quarrel with that. But throughout are casual references to the notion that chimpanzees, gorillas, and perhaps other apes "have become fairly fluent . . . in sign language, . . . certainly seem capable of using language to communicate," and so on. The bonobo Kanzi, of whom we will hear more later in this book, "remembers and describes" a spot in the woods. One of the several books covered in this review, the novel *Jennie*, involves a chimp who is taught sign language, and "learns to express herself."

All of this takes for granted that, with proper training, some nonhuman primates (and perhaps other animals as well) can be provided with the gift of language, even if their species has not yet figured it out. The notion is certainly not unique to this reviewer, Douglas Chadwick: a 1996 novel, *The Woman and the Ape* by Peter Høeg (the author of *Smilla's Sense of Snow*), involves an ape who is brought to language school. The wife of the experimenter comes to feel that the ape is being exploited. She takes up with him, and they run off together to have an extremely expressive relationship.

Many readers will recall George Orwell's classic *Animal Farm,* where the animal characters are fully fluent in English: they even manipulate one another by manipulating the language. When I was a child, I read a series of "Freddy the Pig" books (by Walter R. Brooks) that also involve a barnyard full of talking animals. While Orwell's book is allegorical, and I did not take Freddy and his colleagues all that seriously, Chadwick's review and Høeg's novel are not meant to be allegory, childish fantasy, or science fiction. Presented as having a basis in current science, they are intended as novelistic treatments of possible situations. Chadwick certainly thinks, for instance, that the author of *Jennie* "seems thoroughly versed in ape research and in the debates surrounding it, and for readers unfamiliar with the subject, his well-intentioned novel makes a fine introduction."

To the extent that Chadwick's assessment is shared, the ability of suitably trained apes to converse with us in a natural language (at least with proper training) has become a more or less accepted fact. It gets worse: as the article shown in Figure 1.1 makes clear, the vanishing distinction between the abilities of humans and of other primates to use language may even be something for naive Web surfers to worry about . . .

Yet, as Chadwick puts it, a proper appreciation of animals' cognitive capacities in this domain is threatened by a band of unsympathetic characters who are "intent on preserving language and reason for the exclusive use of humans." These are the so-called linguistics experts—folks such as the present author. Intent on defending the exclusivity of our scientific turf, we comprise curmudgeons, romantics, and/or elitists who cling to human uniqueness with respect to language in the face of the apparent facts.

Actually, as David Pesetsky pointed out in his response to Chadwick's review, published in a later issue of the *New York Times Book Review,* linguists would be "delighted and intrigued to discover" language in the relevant sense in other primates—or in cockroaches, for that matter. When we look closely, however (and experimenters have tried *awfully* hard), that is not what we find. It appears to be an empirical result, not merely an anthropocentric prejudice, that human language is *uniquely* human, just as many

Figure 1.1 Vanishing distinctions bring new threats

complex behaviors of other species are uniquely theirs. Doctor Dolittle, despite his good intentions, was laboring under a misapprehension.

Chadwick's review inverts the usual logic of the literature about the behavioral and cognitive abilities of animals. What we more often hear is that "apes (chimpanzees, gorillas, . . .) are a lot like us. Therefore, there is no reason in principle why they could not control a language, just as we do."

Chadwick's argument goes the other way: he suggests that since apes really can express themselves and communicate in a language, they must be a lot like us; therefore we should be more considerate of them. Surely, though, we do not need this argument to arrive at the conclusion that considerate and humane treatment of animals is warranted. It is a good thing we do not, because when we look at the evidence, there *do* seem to be significant differences in the language-using abilities of humans and other apes.

Of course, we do have much in common, and it is meaningful to study and understand these commonalities. Their existence, though, does not mean we have (or could have) *everything* in common. For instance, no one denies that humans and bats share a great deal by virtue of being mammals. But even the most dedicated and brightest of human children could hardly be trained to fly by vigorously moving their arms about, or to use echolocation to catch insects. That we are clever enough to build airplanes and sonar systems to accomplish similar ends in different ways does not alter this fact: there are genetically determined differences between humans and bats that establish the limits and possibilities for each.

It seems likely that the human capacity for learning, speaking, and understanding languages is determined by our innate cognitive and neural organization, and as such is uniquely accessible to organisms that have the same specific organization. This capacity develops in the course of human maturation, in the presence of relevant experience—much as other cognitive systems, such as vision, have been shown to do in more limited ways. In the absence of the appropriate biologically based organization, the experience that gives rise to our knowledge of language cannot have that effect, no matter how carefully structured.

Aha, you say, the bat analogy misrepresents the issue. We can't fly because we don't have wings, and we can't catch bugs for lack of the right sensory organs for echolocation. Since language is a kind of behavior, not a physical organ, the argument from genetics fails. Humans and, say, chimpanzees both have brains, mouths, and ears, and those brains, mouths, and ears are quite analogous in their overall structure. Furthermore, humans do not develop language uniformly, the way bats of a given species all come to catch bugs the same way. Rather, we each learn the particular language that happens to be spoken by the community surrounding us; surely that proves that language could not be innate.

But consider this estimate cited by Steven Pinker: "Half of our 100,000 genes are expressed primarily in the brain, [and certainly] species differ from one another innately, [and] humans differ from one another innately on every quantitative trait, and . . . human cognitive accomplishments are solutions to remarkably difficult engineering problems, [so] I myself don't doubt that much of neural organization is innate. Of course that leaves open the question of what aspects of language in particular are innate." With the recent mapping of the human genome, we now know that the actual number of genes is probably less than half the number Pinker cites. Nonetheless, the estimate of the proportion of genetic material devoted to the brain and nervous system continues to "range from 'a fair chunk' to '40%' to 'most.'"

There are excellent reasons to see much of behavior and cognition as closely related to the genetically determined organization of the organism, and thus at least adequate reasons to speak of a human language "organ," with a structure determined by human genetics. Organisms with this organ acquire and use languages of the human sort, whereas organisms without it do not (and cannot), any more than we can fly or catch mosquitos by echolocation in the absence of the relevant species-specific equipment.

How much of language is determined by our uniquely human genetics? To address the question, I need to clarify what we mean by *language*. This goal, in turn, requires distinguishing a specific sense of *language* from a much more general sense that is close to the broad notion of *communication*.

We commonly talk about all sorts of things as language—the language of dreams or of films, body language, even the language of traffic lights. Common to all of these is that they involve communication: one individual (or the film, or the traffic light) emits some kind of signal from which other individuals can derive information. Surely it is not *that* sense of language which is at stake. Everyone grants that organisms a lot less complex than chimpanzees communicate. We would not want to say, though—because organisms of all sorts can determine information from olfactory, visual, or other signs about when an individual of the opposite sex is interested in mating—that no fundamental distinctions can be made, and that language is really universal. The issue is not whether communication takes place in all these circumstances, but rather *how* that communication takes place, and what sort of system it is based on. When we make these inquiries about human communication, a rather special and much more specific sense of "language" emerges.

What I am talking about, more specifically, is the use of systems such as English, French, Japanese, or Potawatomi. Just what *is* a natural language? The definition is at bottom what linguistics is all about, and any snappy, aphoristic definition is virtually bound to fail. In general, every science starts from a presystematic notion of its subject matter, and its results serve to provide a more systematic reconstruction of the properties of the object of inquiry: rocks, molecules, organisms, political systems and economies—or languages. If we could sum up the significant aspects of any of these items in a few sentences, the scientists who study them could leave for the beach, their labors complete.

Short of a completed science, though, treating natural language the way the U.S. Supreme Court has sometimes treated pornography ("I know it when I see it") moves us quite a distance. We know that English, French, and others are natural languages in ways that traffic lights or cinematic symbolism or Fortran, for example, are not. We may not always know what *a* language is (witness the Ebonics discussion of the late 1990s), or when one language is the same as another (consider the sense of "Serbian" as opposed to "Croatian" or "Bosnian," three largely similar forms of what used to be called "Serbo-Croatian," before the breakup of the former Yugoslavia in 1991–92). Nonetheless, we know there is a difference between *language* and other forms of communication.

For generations, philosophers have agreed that the remarkable feature that gives human language its power and its centrality in our life is the capacity to articulate a range of novel expressions, thoughts, and ideas, bounded only by our imagination. Using our native language, we can produce and understand sentences we have never encountered before, in ways that are appropriate to entirely novel circumstances. We will see in Chapter 8 that human languages have the property of including such a discrete infinity of distinct sentences because they are *hierarchical* and *recursive*. That is, the words of a sentence are not just strung out one after another, but are organized into phrases, which themselves can be constituents of larger phrases of the same type or other types, and so on without any boundary.

It is this structural property that gives language its expressive power, so it is reasonable to ask of any candidate for comparable status that it display recursiveness as well. We will see that there is much more to the characteristic syntactic structure of human languages than just recursion, but this is incontestably a core property, sine qua non.

The central issue of this book comes down to a pair of related questions. To what extent do animal communication systems share essential properties with those of human language? (For the reasons just described, pay particular attention to the question of whether these systems display the characteristic properties of unboundedness, hierarchical organization, and recursion.) And if there do indeed remain significant areas of nonoverlap, can any animals other than humans be *taught* to use a communication system with the essential properties of a human natural language?

These questions define my agenda here: to arrive at an understanding of the way animals communicate in nature, to show how the properties of animal communication systems relate to those of human natural languages, and to determine whether the differences we find can be bridged by training. In the process I survey a number of different animal systems, and also provide enough of an introduction to the characteristics linguists have found in human languages to make the comparisons scientifically meaningful.

Chapter 2 begins by discussing briefly what "communication" is, together with attempts to define language in terms of a set of necessary and sufficient conditions on communication systems more generally. Checklists of this sort invariably end by misrepresenting the object they attempt to characterize, and cannot substitute for a more detailed and nuanced exploration of its properties.

Chapter 3 addresses two sides of a basic problem in studying cognition. In some instances we tend to overinterpret what seems complex to us, while in others we take too much for granted about behavior that appears simple and straightforward. I discuss some of the classic pitfalls in trying to answer questions about animal cognition. If we want to be neither tryingly skeptical nor irrationally exuberant about animals' abilities, subtle questions must be taken into account in interpreting their behavior, especially when that behavior seems strikingly flexible and appropriate to a situation as we interpret it. The other side of this coin is the likelihood that the apparent simplicity and ease with which we deploy our own skills as language users belies the complexity of the system involved, a complexity rooted in human biology. Neither the fundamental intricacy of a behavioral pattern nor its essential simplicity can necessarily be read from its immediate appearance.

Continuing the exploration of the way one investigates cognition, especially in nonhumans, I turn in Chapter 4 to one of the best-known examples in the animal communication literature, the dances performed by forager honeybees. These dances provide information that fellow bees *could* use to locate the desiderata of apian life: pollen, nectar, and potential locations for new colonies. However, that fact alone does not determine the correct interpretation of the dance behavior. In the process of studying this system, I make another methodological point: a good story is not necessarily self-validating, although in the end it may turn out to be true.

I then touch on matters more specifically related to the nature of language. In Chapter 5 I discuss some fundamental properties of sound, the medium in which most linguistic transactions occur. Understanding the acoustical structure of the sounds organisms produce, how they produce those sounds, and how sounds are dealt with by the brain and the auditory system is essential to any account of communicative behavior. I begin with a system that is comparatively simple, the calls of frogs. The frog's production and perception systems are closely attuned, making the animal especially adapted to respond appropriately to the specific sounds that are ecologically important to it. This lesson is applicable to a broader understanding of perception, including the analysis of speech in humans that occupies the bulk of the chapter.

In Chapter 6 I look at an even more elaborate acoustic system, that of birds (especially of oscine songbirds). Interesting parallels exist with some properties of human language, though many fundamental differences are present as well. One intriguing possibility that emerges from the study of birds is that of tracing connections between the systems of song production and perception in much greater neurological detail than can be done in other organisms, suggesting conclusions that dovetail nicely with proposals about human speech. Again, biologically determined systems that specialize in the processing of ecologically important signals emerge. The most significant human parallel, however, is probably with the development of a bird's song system, an area that has been the object of enormous research. Similarities between the acquisition of song by birds and of speech patterns by human infants are strong enough to merit a fairly extended discussion.

Primates are the focus of Chapter 7, where I consider some of our knowledge about the communicative behavior of prosimians, monkeys, and apes in nature. This discussion centers on the set of alarm calls that a variety of primates produce in the presence of predators. These raise important questions about the extent to which we should ascribe meaning to animal signals in the sense that words of a human language refer to objects in the world external to the speaker. Besides alarm calls, primates produce a variety of other vocalizations that have communicative importance. We can learn from these calls, but the range of their external expressions turns out to be rather restricted. If writers pessimistic about the mental life of non-humans are to be believed, the animals might just have very little to say—but the evidence for sophisticated thought processes is hardly negligible.

What does account for the massive differences in expressive capacity between human languages and the communicative systems of other animals? As already suggested, the answer turns out to be a central (if often misunderstood) property of language: the system of syntax, with its hierarchical and recursive structure. For those whose only systematic exposure to grammatical analysis came in high school English classes, syntax may seem only a perverse, prescriptive fixation. That is not at all the case. In Chapter 8 I sketch a few of the remarkable syntactic properties of human language, and some of the reasons to believe that this organization is a genetically determined capacity specific to our species.

In Chapter 9 I build a foundation for addressing another of the questions posed above, concerning efforts to teach our languages to other species. To this end, a consideration of the properties of manual (or signed) languages is in order. These have been the basis of the best-known and most ambitious experiments of this sort to date. Contrary to popular opinion (including that of the cat's-meat man quoted at the start of Chapter 9), science has shown that manual languages such as American Sign Language, or ASL, have all the essential structural characteristics of natural languages such as English or Arabic, even though they involve gestures other than those of speech.

If an ape really could learn to use ASL, that would count as learning a natural language. It was in that direction that researchers concentrated their efforts in the 1960s and 1970s. I survey a number of those projects (Washoe, Nim, Koko, Chantek) in Chapter 10, along with other studies that abandoned all similarity to the actual modality of human natural language (speech or sign) in favor of purely arbitrary symbols played out on a keyboard or plastic tokens arranged on a board. For a variety of reasons, all fall far short of demonstrating language abilities in other species.

The most interesting—and also the most scientific—work of this sort that has been done involves apes of a different species, bonobos (often misleadingly called pygmy chimpanzees), and particularly the justly celebrated Kanzi. These animals appear to come somewhat closer than other apes to what we might call genuine linguistic ability. Kanzi's interpretation of certain spoken English sentences is particularly seductive. The ultimate conclusion nonetheless seems to be that when we look at the parts of the system apes can learn to control, the crucial distinguishing properties of language (especially recursive syntax) are still missing.

It is worth stressing once more that this negative conclusion is not the reflection of some presumed species-centrism on the part of linguistic science. If we were to find that other species (say, bonobos) could truly learn the significant parts of a human language, the result would fascinate linguists, not repel them. On the available evidence, though, no such claim seems warranted.

Short of actually learning a language, some of the animals in these studies have demonstrated abilities involving the use of arbitrary symbols for rather abstract concepts. Were we to think of language exclusively in terms of symbolic communication, that would suffice. The actual richness of the expressive capacity of human language, though, depends on further elaborations of exactly the sort that animals do not achieve. Exploring the abilities they display in these studies (but not, apparently, in nature) is certainly relevant; but that is a separate issue from whether or not they have the capacity to learn and use a language in the specific sense that refers to human languages.

By using the expression "human language" repeatedly, I do not of course mean to exclude a priori anything a nonhuman might do. The properties of language that I discuss in the chapters to come are abstract enough to be dissociable from the activities of human vocal tracts and ears, hands, and eyes. They would be directly identifiable in the behavior of other animals if they were indeed found there. Nothing about language in the sense intended here is intrinsically limited to systems with our specific physical organization—though as a matter of empirical fact, the capacity for language *does* seem to be limited to organisms with our specific neurological and cognitive organization.

Research that has been conducted with an African grey parrot named Alex supplies a cautionary note concerning our lack of success in teaching human language to animals. Alex does some remarkable things, more impressive in many ways than the linguistic accomplishments of the widely touted chimpanzees. That should give us pause in interpreting the research done with primates, because common sense would seem to tell us that chimpanzees are smarter than parrots. Still, even Irene Pepperberg's fascinating work falls short of what it would take to demonstrate a capacity for something with the essential properties of human language in another animal species.

On the basis of the available evidence, language as it appears in humans seems inescapably to be a uniquely human faculty with its own unique characteristics, part of the biological nature of our species. If that is the case, language must have arisen in the course of our evolution, separate from that of other primates. In Chapter 11 I survey the little that is known about the precise course of those developments. In the end, I return to the conclusion that the distinctly human ability that has arisen in us is not, as often assumed, the capacity to use arbitrary meaningful symbols, but rather the ability to combine those symbols syntactically.

I do not discuss many of the other animals whose communicative abilities have been the object of various studies. We do not really know what the structure of such systems might be. For instance, the communicative behavior of elephants has evoked a good deal of interest in both scientists and the general public. It seems likely that elephants produce very low frequency sounds with considerable energy, acoustic waves that can be detected by other elephants at great distances. There is little doubt that listening elephants can derive information from these sounds, and that this may influence their behavior. So is there a language of the elephants? In the sense of a language of traffic lights, obviously there is. But until we have some understanding of just what messages this system can convey, what aspects of the signal's structure are relevant to determining those messages, or how much the elephants "sending" the message intend to communicate some particular meaning, we cannot say much about it.

The same is true of other, even more famous cases, such as the vocalizations of whales and dolphins. It is abundantly clear that these animals are highly intelligent and that they engage in communicative behavior. The structure of their communications, however, is simply not understood. While it would certainly be interesting if it turned out that way, there is no reason beyond wishful thinking to believe that when we do come to understand the nature of cetacean vocalizations, they will have the essential structural properties of human languages—whatever fascinating specific properties they *do* have.

Part of my intention is to convey a sense of the remarkable diversity of the species-specific means of communication used by the world's animals. If some of the irreducible particularity we find has the consequence of setting off human language from other systems, that is no more surprising than the discovery that other biological specializations lead to equally particular, indeed unique, abilities in many animals. These differences are not a matter of philosophy, theology, or misplaced humanist sympathies; they are empirical features of nature. We may not be able to take flight by flapping our upper extremities, but we are the only species known that can rationally discuss our inability to do so. As Bertrand Russell famously put it, "A dog cannot relate his autobiography; however eloquently he may bark, he cannot tell you that his parents were honest though poor."

2

Language and Communication

At tea-time, when the dog, Jip, came in, the parrot said to the Doctor, "See, HE's talking to you."

"Looks to me as though he were scratching his ear," said the Doctor.

"But animals don't always speak with their mouths," said the parrot in a high voice, raising her eyebrows. "They talk with their ears, with their feet, with their tails—with everything. Sometimes they don't WANT to make a noise. Do you see now the way he's twitching up one side of his nose?"

"What's that mean?" asked the Doctor.

"That means, 'Can't you see that it has stopped raining?'" Polynesia answered. "He is asking you a question. Dogs nearly always use their noses for asking questions."

—The Story of Doctor Dolittle

Communication is virtually universal among living things. Even bacteria communicate. Some classes of bacteria secrete distinctive organic molecules, for which they have specialized receptors. This apparatus allows the bacteria to detect the presence of others of the same species, a system known in the literature as quorum sensing. "Bacteria, it turns out, are like bullies who will not fight unless they are backed up by their gang. An attack by a small number of bacteria would only alert the host's immune system to knock them out. So bacteria try to stay under the radar until their numbers are enough to fight the immune system." The molecules secreted by one bacterium serve to communicate its presence to the others. Yet surely not all communication is of a piece with all other communication: the use of the word *talk* in the title of the *New York Times* story about quorum sensing is simply the journalist's effort to be clever.

To determine the true issue here, consider an example. One evening I returned home to find my wife correcting papers for her French class. When I asked her what we were doing for dinner, she said, "I want to go out." That is, she produced a certain sequence of sounds, and as a result I knew that she wanted us to get in the car and drive to a restaurant, where we would have dinner.

When I came home the following night, I found my cat in the kitchen. She looked at me, walked over to an oriental rug in the next room, and began to sharpen her claws on it. She knows I hate that . . . and as I came after her, she ran to the sliding glass door that leads outside. I yelled at her, but my wife said, "Don't get mad; she's just saying, 'I want to go out.'"

We conclude that both my wife and my cat can say "I want to go out." Do we want to assert that they both have language? Surely that is at best an oversimplification, although it is clear that both can communicate. Each can behave in such a way as to convey (somewhat similar) information to me.

Here is a sketch of how "real" communication takes place: One organism has a message in mind that he or she wants to communicate to another organism. He or she emits some behavior (makes a noise, scratches the carpet) that encodes that message. The other organism (me, for example) per-

ceives the behavior, identifies it in terms of the meaning encoded, and treats the result of that decoding as the meaning of the message.

Sometimes called the Message Model of Communication, this description may seem fairly obvious, but is it a valid general definition of communication? Communication can take place even when there is no evident basis for saying the communicator "intends" to communicate anything. Think of our bacteria above, or a blush, or the visible signs in many species when a female is in estrus and receptive to mating: there is no intention on the part of the signaler, but a message is communicated all the same.

On the other side, it may be that the recipient interprets the message only in part on the basis of its literal content and relies also on various non-overt contextual or social factors. Consider "Can you pass me the salt?" Here the literal content is an inquiry about the listener's physical capacity to perform an action, but the message usually conveyed is a request that the salt indeed be passed. Or perhaps I ask my colleague what she thinks of the candidate we have just interviewed for a job, and she says "He seems very diligent." In an academic context, this implies a very negative recommendation. If a candidate's best quality is diligence, it is *not* creativity, imagination, or inspirational teaching. In both examples, clearly the linguistic content of what we say may be quite different from what we communicate.

The little story about my wife and my cat illustrates the characteristics of any communication system. First, what is the nature of the behavior or other signal? The cat scratches the carpet and runs to the door to convey a message we might interpret as similar to one my wife conveys by moving her vocal organs to produce sound. Second, what is the range of messages the system can convey? Evidently, my cat can say fewer things than my wife: what is the basis of this difference in expressivity? Third, what relation, if any, is there between the message expressed and the communicator's intentions? The cat certainly intends *something*, but her behavior actually reflects her internal state; my wife can say what she does even if she doesn't really want to go out. Finally, what is involved on the receiving end? Obviously, you have to know the code in order to get the message, but what else? My wife and I understand the cat's scratching behavior as attention seeking in the context of my evident and constant displeasure at it, but is there some kind of underlying code that all three of us share?

Another important aspect of communication systems (not significant in this case) is how the communication system came into being. Did it evolve gradually out of something else, or did it spring into operation fully formed? My cat scratches the carpet basically to sharpen her claws; whatever additional meaning may accrue to that action has grown up ad hoc between us. Most systematic means of communication have more interesting and far longer histories.

This is an area of inquiry where the questions that can be raised are potentially more interesting than the answers currently available. Historical evidence for the sounds of language is minimal; even the soft tissue of tongues, ears, and brains leaves no trace in the fossil record.

The original nineteenth-century constitution of the Société linguistique de Paris is famous for explicitly prohibiting the discussion of matters concerning the origin of language at the society's meetings. This was no mere quirk of the founders: they introduced this limitation for precisely the reason that there could apparently *be* no real science that bore on the topic. Since the late 1990s, interest among linguists and others has reawakened, and conferences are now regularly devoted to the subject. To my mind, this revival is not based on additional data, but rather on the mistaken impression that if we can *pose* an important question, we ought in principle to be able to find an answer. Fortunately, we need not resolve this vexatious problem before studying communication systems and communicative abilities *comparatively* across animal species. We will return to these matters in Chapter 11.

Notions of Language and Communication

How might we distinguish between "language" and "communication"? One way of approaching the distinction is to note that communication is something we *do*, whereas language is a *tool* we can use. We can, of course, communicate without language, though the range of material we can transmit is limited in significant ways. Most of the amusement value of the game of charades, for instance, lies in trying to circumvent these limitations. In fact, a desirable skill in this game consists in referring to words without actually using them (using gestures interpreted as "short word," "sounds like," and so on).

For comparison, the activity of building houses is also something we do, and we use particular tools to do it. Without hammers, nails, saws, and levels, we could not practice the construction trade as we know it. Yet that does not mean we could not construct shelters. We can do a certain amount of building without tools, or using different tools, as other societies do. Still, the structure of the tools makes certain sorts of construction easy and natural. We can study the structure of the hammers and saws and ask where they come from. We see, of course, that there is a close connection between the structure of the tools and what we can do with them, but we should not confuse the activity of carpentry or construction with the tools we use in pursuing it.

Suppose we want to open a nut. We do it by exerting force on the shell through a hard object—either with leverage, using a nutcracker, or by hitting it, for instance with a hammer. Chimpanzees in the wild open nuts by putting them on one rock, then hitting them with another rock—a technique similar to one used by humans. The tools are not identical, but they have the same structure in the relevant respects. There is an activity, and similar means are used in carrying it out. As far as communication is concerned, we do a lot with facial expressions, grunts, and the like. Again, considerable similarity among human and nonhuman primates exists in the activity and in the means for executing it.

Orangutans in nature do not use tools equivalent to those of human carpentry. But if we give an orangutan a claw hammer, and he knows that something good to eat is inside a wooden box that is nailed shut, he can use the claw hammer to remove the nails and open the box, much as a human would. Provide him with the tool, and his cognitive abilities are certainly adequate for using it in some of the ways humans do—ways that depend on the essential structure of the tool.

I imagine that chimpanzees can learn fairly quickly to open nuts with a nutcracker by utilizing the structure of the tool, which is novel to them but suited in form to the task. Yet if we give a chimpanzee a small tape recorder, I seriously doubt that the ape could use it to record grunts and send them to be replayed for another chimpanzee in order to communicate a message. The principal use of a tape recorder might be to serve as the base on which to put a nut in order to smash it.

These distinctions are important when asking whether another species (say, monkeys or apes) can use language. Provided with the proper tools, an ape can use them to engage in at least some "carpentry." What about language and communication? When we ask whether animals other than humans can engage in communication, the answer is, obviously. What is the structure of the means they use to that end, and how closely does their communicative activity resemble human natural language? If we supplied an ape with a human natural language, how much communication could he or she achieve? We need to know a certain amount about the structure of human natural language if we are to make these questions precise; the more we know, the more precise we can be.

In nature, the range of ways in which animals (especially other primates) communicate with one another is certainly not limited to vocalization. Smell (particular substances, such as those secreted by specialized scent glands in both lemurs and rhesus monkeys, as well as normal smells), sight (facial expression, posture), and touching (grooming behavior), among other modalities, also supply information, sometimes intentionally on the part of the communicator and sometimes not.

In terms of the structure of the tools involved, none of these systems seem to fall within a range that might usefully be compared to language. Signals in media such as smell and touch typically are individually simple (that is, they lack a relevant internal structure such that parts of the signal correspond to distinct parts of the message), and in some cases (especially olfactory communication) they are not very flexible in their temporal pattern. There are exceptions: the chemical signals produced and perceived by lemurs may include substances from multiple individuals, deposited at different times; the animals are apparently sensitive to this complexity. Chemical signals in the insect world can be even more complicated. But even where some internal organization is present in the signal, these systems appear to be rather different from human languages.

Characteristics of Language

Now when I speak of "talk" between animals and myself, you who read this must understand that I do not always mean the usual kind of talk between persons. Animal "talk" is very different. For instance, you don't only use the mouth for speaking. Dogs use the tail, twitchings of the nose, movements of the ears, heavy breathing—all sorts of

things—to make one another understand what they want. Of course, the Doctor and I had no tails of our own to swing around. So we used the tails of our coats instead. Dogs are very clever; they quickly caught on to what a man meant to say when he wagged his coat-tail.
—*Doctor Dolittle and the Secret Lake*

What are the essential properties of human language, and how does it differ from other communications? On the face of it, it ought to be possible to generate criteria that would make the difference clear. It turns out, though, that this task is harder than it appears.

One well-known attempt to specify just what properties define a "language" in the human sense was made by the linguist Charles Hockett in the 1950s. Some of Hockett's Design Features of Language may also be found in nonhuman communication systems, but he argued that the whole set is found together only in human languages. Ultimately, the effort to define language in this way poses more problems than it solves, but at least it provides a basis for discussion.

Vocal-Auditory Channel

Language is expressed and perceived in sound. This property is not unique to human language: many species use auditory signals to communicate. Hockett suggests, though, that the particular acoustic spectral features that differentiate messages in human languages really are unique. Vowel color (the property that distinguishes the vowels of *beet, bat, boot* from one another), for instance, is a characteristic of acoustic signals that does not appear to be exploited by the system of any other animal. That is not just coincidental: the development of a vocal tract capable of making distinctions of vowel color is one of the physical specializations for speech that appeared in the course of human evolution. This point is developed at some length in various works by Philip Lieberman and his colleagues.

Just as humans are not the exclusive users of sound for communicative purposes, the vocal-auditory channel is not the only one in which communication occurs. Many others are used as well: signals can be visual, tactile, olfactory, chemical, electrical, . . . In fact, humans communicate with one another (intentionally or not) in most of these ways, though we do not usually confuse that communication with language.

In part because of this diversity, it is vital to distinguish one system from another, so that we speak of a single species as employing multiple communication systems rather than just as "communicating." Each system has its own internal coherence, which can be studied independently of the other systems. Sometimes a single behavior may involve multiple systems in complementary ways. Think of the role played by facial expression in understanding the messages conveyed by accompanying language. The messages conveyed by facial expressions can be explored systematically in one way, language in another. The totality of the communication results from both taken together, but nothing is gained (and coherence is lost) if we attempt to study both at the same time by the same methods.

Indeed, the same channel can convey information from more than one system at the same time. The pitch of the voice on individual syllables serves in many languages (such as those of China and most of the languages of Africa, known as *tone languages*) to distinguish words from one another, in the same way that the difference between one vowel or consonant and another serves this function. Voice pitch is also an aspect of the expressive system of paralanguage (about which I will say more in Chapter 4), which conveys a variety of information about our emotional state, attitude, and so on. The fact that a speaker of Cantonese distinguishes words by tone does not alter the fact that other aspects of the overall contour of that speaker's voice quality and pitch serve simultaneously as paralinguistic cues to excitement or boredom, contempt or admiration.

Separating language from paralanguage is critical to achieving a coherent understanding of the way both systems work. Every time we say something, we communicate much more (or sometimes less) than the literal content. Paralinguistic features (pitch range, loudness, breathiness) have very different properties from sentence structure and meaning. The way pitch is used paralinguistically is inseparable from the way it is used linguistically, even though the two are quite distinct logically.

Hockett took it as self-evident that true language is executed in the vocal-auditory medium. By this he meant to distinguish spoken language especially from writing, which he regarded (correctly) as secondary and parasitic on the spoken language. In fact, not all language (even disregarding writing) *does* involve the vocal-auditory channel. The signed languages used and acquired natively among hearing-impaired individuals have the structural properties of a language such as Chinese or Kiswahili, although they involve the visual channel. Understanding of the richness of the structure of signed languages, and their basic similarities to spoken languages, did not really develop (among linguists, at least) until after Hockett's paper appeared. Chapter 9 is devoted to the properties of these languages.

Broadcast Transmission, Directional Reception

Signals travel generally to any potential receiver, and their properties can help to determine the location of the originating source. This characteristic seems at first glance to apply to just about any communication system, if we disregard the fact that communication can take place even when we cannot locate the source. Think of a disembodied voice backstage in a play, for instance. Of course, in some cases the nature of the medium makes broadcast transmission rather narrow. One example is tactile signals, such as tapping a dancer on the shoulder to indicate a desire to cut in. But not all communication shares this property even in a limited way. Consider the marking of an animal's territory by olfactory signals. At the time the communication takes place—when another animal perceives the scent—the source is not necessarily present, but the signal nevertheless plays its role perfectly well.

Rapid Fading

Many kinds of communication are transitory, in the sense that the signal is not available for inspection for very long after it is produced. Even if you are in a cave with a remarkably persistent echo, the sound fades away within a few seconds. The same is true of signed language or any system of visually perceived gestures; there is not even any obvious analogue of an echo. For logical purposes we can disregard the modern possibility of recording sound or images for later playback, as well as that of writing down what we have heard. Under those circumstances we could consult the transcribed record at leisure, but these special cases are in no way intrinsic to the way speech (or sign) communication works. In this way the modalities of natural languages (both spoken and signed) differ from those of chemical or olfactory signals, or from outwardly visible physical changes that communicate one animal's internal state to another.

Interchangeability

Competent language users both produce and comprehend the same range of signals, at least under normal conditions and barring pathology (such as deafness or blindness). In this respect human language is different from some other systems. Birdsong, for instance, is typically produced by the male and not by the female (although there are species in which both sexes sing) and comprehended by male and female (but in different ways). The honeybee dance is interchangeable among the workers in that these bees both dance and understand the dances of others. The situation is different for queens and drones, who do not dance but do understand the dances—at least to the extent that these indicate possible sites for a new hive and not the location of a food source.

This connection between production and perception of the signals may seem adventitious, but it turns out to be rather important, at least according to some theories. The motor theory of speech perception claims that the way we perceive the speech of others involves a direct reference to what we might have done ourselves. If our perceptual system is truly organized in this way, the fact that hearers are also talkers (and vice versa) is no accident. Humans and songbirds appear to provide strong evidence for this claim.

Total Feedback

Senders can monitor their own signals. Some nonlinguistic signals do not have this property: consider physical changes in an organ outside the range of the individual's vision but visible to others, such as a blush. Blushing certainly conveys information to the viewer, but without a mirror not to the one who blushes.

Some famous instances of communication do not allow for feedback. The stickleback (a fish that was an object of great interest to ethologists in the 1950s) communicates about mating through a characteristic change of color in the belly and eyes of the male and a characteristic distension of the female's belly. But neither can see his or her own belly or eyes.

With respect to human languages, we might feel that feedback, while usually present, is not necessary. Even in the case of deaf speakers or blind signers, though, there is feedback from other modalities (kinesthetic, proprioceptive). Feedback of this sort seems to be significant in learning; both birds and humans (the only well-studied cases of learned communicative behavior) get seriously off the track when it is not available. In normal speech, laboratory conditions that disrupt, distort, or prevent proper feedback can make fluent speech virtually impossible to produce.

Specialization

Hockett's definition of this property refers to the fact that the communicative activity involved in human language does not serve any other func-

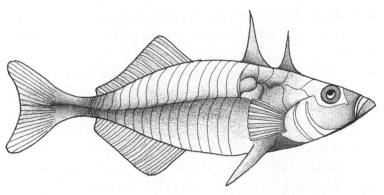

tion: "the direct energetic consequences of an act of communication serve no other biological purpose." Compare this concept, for instance, with the fact that we can derive information from events that *do* have other functions, perhaps more basic than communication. When a dog pants with his tongue hanging out, he is cooling off through evaporation, but he may also be supplying information (especially to other dogs) about the location, state, and identity of the panter. The female stickleback's distended belly, which communicates to the male her readiness to breed, is the result of the development of roe, not of any intent to communicate.

Specialization for communication is tied up with the range of uses for the organs involved. It is sometimes claimed that there are no true speech organs: the organs we use for speech production all have other functions (vegetative, respiratory), and the ears that perceive much besides speech are what we use to hear. The notion that this multifunctionality excludes a specialization for speech overlooks a great deal. In fact, the vocal organs have changed over the course of evolution in the direction of greater functionality in speaking, even at the cost of being less suited to their other tasks.

For example, the natural position of the human larynx is considerably lower in the throat than in other primates, or in mankind's earlier ancestors, or indeed in newborn babies. Among other differences, the tongue is large and rounded, as opposed to the short, flat tongues of other primates; and the vocal tract makes a 90-degree turn, as opposed to the nearly straight vocal tracts of others. Because of its construction, the human vocal tract has a portion that is necessarily involved in the transfer of food and drink, on the one hand, and of air on the other. In our primate relatives (as well as in babies) it is possible to isolate the digestive channel from the respiratory, making it possible to eat or drink and breathe at the same time—but we all know what happens when *we* try that. Evolution, in other words, has modified a nice, serviceable system so as to make it possible for us to choke on our food.

Nonetheless, the resulting system is much more flexible than that of other primates in terms of the range of sound types we can produce. Many basic varieties of vowels and consonants are beyond the articulatory capacity of nonhumans, or of earlier hominids. Evolution has specialized us as speakers. As eaters and drinkers, we simply have to make the best of it.

The perceptual system, as well, seems to have a specialized mode of operation that applies to auditory inputs that have the overall properties of speech. This mode is quite distinct from the one that comes into play in perceiving nonspeech. Under unusual conditions—thoroughly unnatural, but neither impossible nor painful—it is possible to engage both systems with respect to the same stimulus. In the laboratory phenomenon of "duplex perception," we seem to hear both a speech signal (a syllable, such as *ka*) and a nonspeech signal (a sort of falling or rising pitch whistle) in response to the same sound input, when parts of that input are provided to one ear and parts to the other ear. This phenomenon confirms the notion that the human auditory system has indeed evolved a distinctive specialization for dealing with speech, even if we use the same physical ears to hear both the announcer and the crack of the bat when we listen to a baseball game. The idea that there is no speech apparatus per se turns out to be a misconception.

Semanticity

Linguistic forms have denotations: That is, they are associated with features of the world, as opposed to many nonlinguistic signals that refer only to themselves (think once more of the stickleback's belly). I deal further with this issue in Chapter 7, in connection with the meaning of alarm calling behavior. In understanding the workings of language, we want to distinguish semantic signals (which refer to events and objects in the world outside of the signaler) from expressive ones (which simply reflect to the outside world some aspect of the internal state of the subject).

Arbitrariness

It is conventional to observe that the linguistic signal has no necessary relation to what it denotes. Speech signals, that is, are not in general iconic. *Cat* refers in my speech to instances of *Felis domesticus* not because of some perceived resemblance between the sound of the word and some aspect of a real cat, but merely because that is the English word for it. The arbitrariness is reinforced when we observe that other languages have quite different words for the same thing. In Navajo, for example, a cat is a *mósí*, but the cats themselves are just the same.

Arbitrariness is often thought to be falsified in the case of onomatopoeia: thus, a cat says "meow" because . . . well, because that is the noise a cat makes. In fact, though, different languages have at least partially conventionalized onomatopoeic words for animal noises. Cats say "ngeong" in Indonesian, for example. A rooster says "cock-a-doodle-doo" in English, but "cocorico" in French or "kikiriki" in German. A turkey says "gobble, gobble" in English but "glu, glu" in Turkish. A pig says "oink" in English but "groin groin" in French, "röh röh" in Finnish, "chrum chrum" in Polish, "nöff" in Swedish, or "soch, soch" in Welsh. Although these words generally are inspired by sounds made by the animals in question, they are nonetheless words of particular languages, and with very few exceptions they conform to the principles of words in those languages. A pig could not say "groin groin" or "röh röh" in English, because English does not have the nasal vowel [ɛ̃] of the French word or the front rounded [ö] of Finnish. English cats could not mimic their Indonesian counterparts because English words cannot begin with [ŋ] (*ng*), and so on.

As opposed to the words of spoken languages, paralinguistic vocal features are less arbitrary, in that their dimensions tend to be related iconically to those of the internal states they express. Thus, when we are angry, our voice may get loud. When we are angrier, it gets LOUDER—and when we are extremely angry, EXTREMELY LOUD. The dimension of loudness can vary in a continuous way, showing (in principle, at least) as many degrees as does our potential anger or other internal state to which the loudness corresponds. This continuous and iconic character is one of the basic ways in which paralanguage differs from language.

Even apparently transparent iconic communication may have some arbitrariness, though, in the sense that it may have to be acquired in order to be understood. Thus, we take the gesture of pointing for granted as a way to call another's attention to something, but not all cultures use similar gestures in this way.

A story (probably apocryphal) that I heard in an undergraduate class illustrates this point. A missionary is dropped into the jungle and tries to learn the language of the surrounding community. Eventually she learns how to express "What's that?" and sets out to expand her vocabulary. She points to a house and asks "What's that?" and hears "Boogoo-boogoo," so she writes in her notebook: *house:* [bugubugu]. Then she points to a tree and asks "What's that?" and again hears "Boogoo-boogoo." She decides she must have been wrong the first time, and that [bugubugu] means *wood*, not *house*. But she points to a passing dog and asks "What's that?" to which the response is, once again, "Boogoo-boogoo." Eventually she learns that [bugubugu] actually means *right index finger.* In the local culture pointing is done with the chin, and every time she asked "What's that?" her position had been such that her chin was directed toward the pointing finger.

Although various nonhuman primates assuredly have a sense of drawing attention to an object, most do not understand finger-pointing gestures as the way to do so, at least not without extensive training.

Discreteness

The linguistic signal is subdividable into separate units (sounds, syllables, words, phrases . . .), and relatively small inventories of these basic elements can be combined in various ways to generate a much greater variety of messages. As Steven Pinker puts it in *The Language Instinct,* and as I discuss in greater detail in Chapter 8, a language is a *discrete combinatorial system.* Other communication systems have a kind of discreteness—birdsongs are made up of parts, for example—but it is not the same. Some birds can do a certain amount of recombining of the basic elements (analogous to syllables) of their song, but the result is always a variant expression of the same message. The key here is that birds cannot combine syllables in different ways to produce substantively different messages.

The signs used by baseball coaches and managers constitute a different system that also displays discreteness. This system can express a broader array of messages than can birdsongs, but it still lacks the meaningful recombinability we find in language. The coach touches the letters on his uniform twice, then spits (baseball players and coaches do a great deal of spitting), then tickles his right ear, then pulls the lobe . . . and this means *don't swing.* Other combinations mean *hit and run* and the like. The system is based on a set of discrete elements that can be combined in different ways, but each combination is a single unit: that is, there is not some part of the message *hit and run* that is associated, say, with the spitting. In language, on the other hand, the word "hit" in (spoken) "hit and run" is associated with a specific subpart of the total meaning.

This difference is sometimes referred to as that between "phonological" syntax and "semantic" syntax. In a system with phonological syntax, the individual signals have an internal structure and are made up of component parts that can be combined in various ways for different signals or variations on the same signal. The parts themselves do not make discrete contributions to the signal's meaning, however. To the extent that we can find discrete components within internally complex signals, a number of animal and other communication systems can be said to display phonological syntax, but only a system like that of English has semantic syntax.

Animal systems are either discrete or continuous. If they are discrete, they are made up of a small number of possible signals (on the order of five to fifty) that are not semantically recombinable. If they are continuous, different messages correspond to different values on some dimension. The notion of "continuous" here comes from the mathematical sense of the word. It refers to a physical scale (such as direction or distance) with the property that for any two values, there is always (at least in principle) another possible value intermediate between them. The bee dances described in Chapter 4 are examples of a continuous communication system.

Displacement

With language, we can refer to objects and events that are distant in space and time from the location of the speaker or the hearer. Other signaling systems do not in general have this property. To the extent that it makes sense to describe animals' signals as "referring" to something, it is always to the here and now — the attitude or the internal state of the animal doing the signaling.

Even rather rich systems devised and used by humans share this limitation, to the degree that they are not basically parasitic on language itself. A baseball coach may have a signal for *hit and run*, but there is none for *if we're still ahead in the seventh inning, I'll be able to take you out for a pinch hitter.* Bee dances perhaps are an exception, if we think of the bee as "describing" the properties of a distant food source to her fellow workers. Still, it may also make sense to think of this system as reflecting the bee's own internal state, a state that results (here and now) from the foraging flight she has just undertaken. If we think of the hive as both the location of the dance and the origin of the flight vector it indicates, the putative spatial displacement is less evident. In any event, there is no question of a temporally displaced referent: Bees' dances relate to food sources available within a very short temporal horizon from the present, not to where they found a nice patch of hollyhocks last season.

Openneʃʃ and Productivity

An open or productive system is one that is capable of expressing an unbounded range of possible meanings. Most animal communication systems serve to convey at most a few dozen different possible messages. Once again, bee dances are a possible exception. Since the parameters of the dance can potentially distinguish a continuous range of possible food source locations, it follows that the number of distinct "messages" is unlimited, at least in principle.

If we ignore the point that the bees themselves may not be infinitely precise in producing and interpreting these dances, there can thus be an infinite number of dances. This kind of productivity is completely different from what we find in human language. Even on the most charitable interpretation, the bees are always "talking about" the same thing (however many subtly distinct variants there may be), whereas the variety of things humans can refer to when talking is not limited in that way. This difference in the productivity of communication systems requires us to distinguish *continuous* openness — as illustrated by bee dances — and *discrete* openness of the sort we find in natural language.

Duality of Patterning

In every language, units can combine to make larger structures at two quite distinct levels of analysis. Sounds combine to make words, the smallest meaningful units, and words combine to make sentences or whole propositions.

This duality makes for a certain efficiency of language, in that there are not very many different units that have to be kept apart in production and perception. If we needed to learn completely different signals for every word of our language, we would quickly reach the limits of our acoustic and auditory abilities. Compare alphabetic with ideographic writing systems, for example. A child learning to read and write a Chinese language, or the kanji characters necessary to read a Japanese newspaper, can testify to the immense burden of having to memorize all the individual signs separately, as opposed to a child learning the twenty-six signs of the roman alphabet. We avoid that problem by having only a small number of basic sound types (each with no meaning in itself) and combining them in an unlimited number of ways to make larger structures—to which we can assign individual meanings, and then combine *those* into more complex expressions.

Traditional Transmission

A person's language is *learned* rather than completely built in. For example, a child of any genetic background will learn whatever language is spoken in the surrounding community. Certain birdsongs (in three out of twenty-seven orders of birds) are also learned, and possibly also some cetacean vocalizations. In contrast, the (nonsong) calls of birds, primates, and other animals, as well as the broad scope of their applicability, are not learned. As far as we know, all other communicative behavior on the part of nonhumans is innate.

Although most communicative behavior thus is *not* learned, it is not necessarily independent of experience. For instance, some tuning of the appropriate conditions for use of vervet monkey alarm calls apparently takes place during growth and development, although the signals themselves need not be learned. This pattern is characteristic of a great many systems. It is responsible, for instance, for the development of numerous local "dialects": a given vocalization may have a range of possible realizations rather than just one, or a range of possible uses. Individuals may attune their choice from within such a range to the usage of those around them, even though the basic system develops in them without reference to the behavior of others. The claim of innateness in many communicative systems is thus subject to the qualification that the ways in which such behavior is used may be modified somewhat over the life span of the individual.

Prevarication

Some theorists of language origins, fond of paradoxes, have said that language "must have been invented for the purpose of lying." Charitably, we can interpret this statement as emphasizing that language can describe things that are not literally realized or true. We can talk about unicorns and squared circles, even if we cannot ever point them out. We can also use language to lie more literally, referring to states of affairs that are contrary to what we know about the world—doing so not just to exercise our theoretical imaginations, but to actively mislead our listener. Yet insofar as a communicative signal is simply an external manifestation of an animal's internal state, it is not really possible for the animal to "lie."

Some reports suggest that animals other than humans use supposedly communicative behavior to deceive. One celebrated case is that of the (mother) piping plover (*Charadrius melodus*), who apparently pretends to be hurt to distract predators from the nesting site where her relatively helpless offspring would be endangered. But is this undeniably effective strategy really an instance of intentional misleading? The plover's behavior is not just a reflex: the bird clearly tries to lead the intruder away from the nest. I discuss the interpretation of this case in Chapter 3; to anticipate the conclusion, there is no reason to believe that the bird is lying so much as engaging in behavior that she knows will attract a predator away from her nest.

Vervet monkeys sometimes behave in ways that could be seen as an attempt to deceive their fellows about the presence of predators, and Dorothy Cheney and Robert Seyfarth have explored ways to disentangle intended deception from other interpretations. There is also a substantial (and controversial) literature on apparently deceptive behavior in primates, under the heading "Machiavellian intelligence." A substantial corpus of incidents has accumulated, but the evidence remains at the level of intriguing anecdotes rather than systematic patterns of behavior.

To say that some communication is genuinely deceptive, we would want to establish that the communicator has some sense that the recipient of the message has a view of the world, and that the communicator is attempting to manipulate that view (rather than directly manipulating the behavior itself). It is a thorny issue, and one that has been much discussed. Do any animals other than humans have a *theory of mind,* in the sense that they see other individuals not merely as acting but as as holding opinions that underlie their actions? There seems to be no valid evidence for this claim in any species, and some evidence in higher primates that argues against it. For current purposes it suffices to mention that this philosophical question is relevant to the notion of deceptive communication.

The fact that we can use language to talk about things that are not true, or not possible, or simply imaginary, is qualitatively quite different from this point. We can also use language to lie and deceive, but it is difficult to see that as its principal role in our lives.

Reflexiveness

We can use language to talk about language itself. This is a property of human language, and indeed of no other communication system. Birdsong cannot be used to make a comment on another birdsong, for instance, but only to put out a message. The significance of this ability is far from clear, however. Yes, it is possible to write a book such as the present one about the nature of language, but even its author has to admit that this is a fairly marginal activity. To consider it one of the defining properties of human language puts the cart before the horse.

Learnability

Any normal human can learn any one of a variety of languages, depending on the data available in infancy when the person's first language is being formed (or when the language organ is growing). This point seems essentially the same as the one made above about the "traditional transmission" of language, but a subtle difference exists.

When we say that something is learned, that statement can be interpreted in a number of ways. At a minimum, it means that the requisite knowledge or skill developed on the basis of some interaction with the environment. But that is not enough, because the environment alone cannot suffice to explain what is learned. Information in the environment can only lead to learning if the learner is constructed so as to be able to make use of it. And an organism's inherent structure can interact in several different ways with environmental information in order to result in the development of some ability we would call learned.

What we usually have in mind when we talk about learning is what goes on in school: information about facts or methods is presented by a teacher, and as a result the learners develop a more or less accurate representation of the same knowledge or skill. Whatever is learned is, obviously, a direct function of what was taught. Much can be learned in that way, but it is necessary to emphasize that the learning is based on a general intelligence that is quite broadly applicable. And that is not the only kind of learning that we need to recognize.

Paradoxical as it may seem, learning in some instances must be said to take place even though only one possible thing will be learned; different experiences all lead to the same result. All that really matters is that there be *some* relevant experience to trigger learning.

This type of learning is what appears to take place in the lower-level visual system, for example. The visual system is not completely formed at birth. The neural organization that allows for the recognition of specific features develops as neurons form specific synaptic connections with one another over time, as the organism matures. The particular features that emerge in the visual system seem to be essentially uniform across the species, yet the apparatus for recognition can develop only in the presence of visual data. When crucial features of the visual environment are absent from early experience, the system does not develop as it should — even though there is really only one course it can follow. This fact was demonstrated some years ago in classic experiments by David Hubel and Torsten Wiesel on kittens, and much the same point emerges from more recent work on the maturation of the human visual system.

It makes sense to say that we "learn" how to see, then, even though no traditional transmission is involved. In contrast, we apparently do not need to "learn" how to taste. The neural apparatus for detection of the basic taste sensations is completely in place at birth. My colleague Linda Bartoshuk, who works in this area, points out that most of what we call taste is actually a matter of *smell,* or at least of detection and analysis by the olfactory system. Smell, as opposed to taste in the strict sense, is a faculty that involves a lot of learning — and "traditional transmission." We do not need experience to detect sweet, sour, and bitter sensations, but we certainly need to be taught to tell a Bordeaux from a Burgundy.

In the case of language, environmental input is needed in order for the language faculty to develop, although it turns out that, in extremis, infants will seize on remarkably little information as the basis for linguistic maturation. The nature of the system that emerges is strikingly uniform in remarkable ways, and in that sense it is similar to vision; but variations are also present. Different languages have different words, but there is much more than that to say about the range of potential variability. The specific properties of the individual language learned must be fixed in ways contingent on the available evidence.

The learning of language (for a human, at least) is thus intermediate between the kinds of learning discussed above. It is certainly not like the development of the visual system, in that only one possible kind of knowledge can emerge. The specific language a child learns depends on the language spoken in the environment during childhood, not on that spoken by the parents. It is not like learning calculus at school either, because what is learned is not solely a product of what is heard. Human language learning involves a major contribution from the learner. It is more a matter of selecting from a large but limited range of possibilities than of developing an ability from scratch.

Problems with Lists

If we take a set of characteristics such as those just reviewed (perhaps massaged a bit, to allow for possibilities such as signed languages, which Hockett's discussion did not accommodate), we can probably come up with a list of tests with the desired property. All human languages "pass" these tests, and nothing that is not a human language can pass them all. Such lists have tended to define much of the content of current animal communication research. That is, researchers want to establish that such and such an organism either "has language" or at least "has the cognitive ability to acquire language." Their approach is to take a list of characteristics and show that their animal does indeed exhibit all of them, or at least that they can teach it enough to "pass all the tests." Gary Bradshaw calls this the Signature Characteristic Strategy.

Such a strategy raises a number of conceptual problems, as Bradshaw points out. For one thing, any specific set of signature characteristics is liable to need constant revision. For instance, now that linguists understand that manual languages such as ASL have all the structure of other natural languages, we have to rethink the role of sensory modality in the basic nature of language. In general, the tests suggested by Hockett reflect the understanding linguists had of these matters in the 1950s. Much has changed since then — presumably for the better, at least in terms of our understanding of what a language is. Although the cognitive abilities and communication systems of some animal might well "pass" at a given stage in the development of science, the same animal might be said to "fail" one or more tests as the actual battery evolves in accordance with changes in what we know about language.

Researchers exploring the cognitive capacities of other species, especially those involved in ape language, have often complained about what seems to them to be a double standard. Sue Savage-Rumbaugh, a prominent figure in this field, objects that "they keep raising the bar. First the linguists said we had to get our animals to use signs in a symbolic way if we wanted to say they learned language. OK, we did that, and then they said 'No, that's not language, because you don't have syntax.' So we proved our apes could produce some combinations of signs, but the linguists said that wasn't enough syntax, or the right syntax. They'll never agree that we've done enough."

It is certainly true that scientific notions of what constitutes the essence of language have evolved over the years. Contrary to Savage-Rumbaugh's impression, these changes have not been arbitrary and capricious, designed to ensure that nonhuman animals will never be able to succeed. In particular, the centrality of syntactic organization to the expressive power of human language has become much more obvious than it once was.

It is fair to say that in the 1950s, linguists had little understanding of the syntax of human natural languages. Research focused almost exclusively on the systematic role of sound in language and on the structure of words. Only with the development of generative theories of grammar in the 1960s and thereafter have we come to a better appreciation of syntactic structure. That structure is amazingly intricate, and remarkably uniform across languages despite superficial differences. It is undoubtedly based in the biologically determined maturation of human beings, and it contributes in essential ways to linguistic expressivity. If we place more emphasis today on the role of syntax in the evaluation of ape language experiments, it is because we understand language better, not because we have found that this is a way to deny it to chimpanzees, bonobos, and others.

What is essential is recognizing that the evaluation of cognitive research on other species is not some sort of computer game, in which passing a certain number of tests means that you "win." We are trying to understand what animals do and what they do not do, what they can do and what they cannot do. Is it true that they really can learn all of what seem to be the essential structural properties of a human language? Most of these? Only a few? None at all? Whatever the answer to these questions, it constitutes a result, and one in which linguists are interested. If it turns out that either naturally occurring communicative behavior or abilities induced in laboratory experiments indeed have the characteristic properties of a human language, linguists in general will be fascinated, not repelled, by that fact.

Even if it were possible to agree on a more or less definitive set of signature characteristics of language, it might still be difficult to measure the extent to which an organism does or does not exhibit those characteristics. The more what we are looking at diverges from language in its essence, the harder it is to develop meaningful comparisons in matters of detail.

For example, the utterances of young children become progressively longer as they mature. The measure of Mean Length of Utterance (MLU) is presumed to reflect the emerging complexity of the child's linguistic behavior, and it is common to assume that MLU actually measures linguistic maturity. In the ape language experiments, however, the number of signs produced in an "utterance" seems to have quite a different basis; their multiplicity reflects repetition, rather than elaboration of a basic idea. If we assess the linguistic maturity of chimpanzees by measuring MLU, we get the impression that they reach a very high level quite quickly—then get stuck. In actuality, though, MLU measures rather different things in baby humans and baby apes.

If our goal is to understand what a natural language actually is at its core, any checklist of the sort we have been exploring is almost certain to be superficial. A set of criteria that picks out a given class within the world as we know it may not accurately express its essence. Consider the definition of humans as "featherless bipeds." Who would be content with that as an expression of the nature of humanness (or of primateness, if you think of chimpanzees and gorillas as bipeds)?

To understand how human languages are (and are not) related to other forms of communication among animals, we need to do more than devise a set of tests. Chapter 3 explores some of the pitfalls that arise in understanding and appreciating what we find when we study the cognitive capacities of various animals.

3

On Studying
Cognition

Now this Nino was just an ordinary, cream-colored cob who had been trained to answer signals. Blossom had bought him from a Frenchman; and with him he had bought the secret of so-called talking. In his act he didn't talk at all, really. All he did was to stamp his hoof or wag his head a certain number of times to give answers to the questions Blossom asked him in the ring . . . Of course, he didn't know what was being asked of him at all, as a matter of fact. And the way he knew what answers to give was from the signals that Blossom gave to him secretly. When he wanted Nino to say yes, the ringmaster would scratch his left ear; when he wanted him to say no, he would fold his arms, and so on. The secret of all these signals Blossom kept jealously to himself. But of course, the Doctor knew all

about them because Nino had told him how the whole performance was
carried on.

    . . . "Look here, Mr. Blossom," said [Doctor Dolittle] quietly, . . . "I
know a good deal more about animals than you suppose I do. I've given
up the best part of my life to studying them. You advertised that Nino had
understood you and could answer any questions you put to him. You and
I know that's not so, don't we? The trick was done by a system of signals.
But it took the public in. Now I'm going to tell you a secret of my own
which I don't boast about because nobody would believe me if I did. I can
talk to horses in their own language and understand them when they talk
back to me."

                                          —*Doctor Dolittle's Circus*

Much of what looks enormously complex to us in the world is in fact based
on remarkably simple principles. A standard example in complexity theory
is the elaborate constructions termites produce in building their mounds:
these prove in the end to be based on nothing more complicated than each
termite's putting a new bit of the nest right next to the bit added by the
previous termite. Conversely, much that looks simple can be accounted
for only in terms of complicated models. The baseball player's apparently
effortless glide toward a dropping fly ball involves the (virtual) solution of
systems of differential equations based on keeping constant certain angles
in the visual field and other visual properties.

    Because of this disparity between the actual complexity of the world
and our interpretation of it, we must always be skeptical of the extent to
which our first impressions correspond to the way nature really works. This
cautionary lesson is the point of the present chapter, approached in terms
of two apparently disparate topics. On the one hand, we consider some instances of animal behavior that seem "obviously" to reveal elaborate underlying cognitive processes, but that actually lend themselves to much more
conservative interpretation. On the other hand, we take a first look at some
properties of human natural language that turn out to be far less simple
than they appear.

    These two matters may seem unrelated, but to understand the issues
in comparing animal communication and human language it is essential to
appreciate the tension between apparent and real complexity of structure
in behavior and cognition. In evaluating the evidence for language-related
abilities in nonhuman animals, we must be constantly aware of the tempta-

tions both to overestimate what the animals are doing and to underestimate what we ourselves know and do as speakers of our native tongue.

There is another danger. If we measure everything we find in animals by the standard of its approximation to something we might find in humans, we run the risk of underestimating the animals and thereby missing what is interesting, complex, and significant about these creatures' abilities and actions in their own right. We can hardly hope to evade these difficulties altogether, but preliminary study of some examples for their methodological value may be worthwhile.

Giving Animals (Exactly) Their Due

In Chapter 2 we looked at Hockett's classic attempt to arrive at a precise characterization of the difference between human language and other communication systems. His 1960 paper proposed an extensive set of criteria for assessing a communication system, with a specific set of values on these dimensions taken to characterize human natural language. Looking through the other end of the telescope, we could see these tests as a set of criteria such that if, in exploring some particular animal's behavior, we were to find the same set of values as Hockett, then we should attribute humanlike language ability to the animal.

Surely, however, this approach gets things the wrong way around. Even if, in the classic philosopher's illustration, the members of the species *Homo sapiens* are the world's only featherless bipeds, we do not want to equate our humanity with erect posture and lack of plumage. Similarly, the nature of language cannot be reduced to some collection of its external properties, although of course those properties may, if well chosen, tell us much about what we ought to be paying attention to. And if the cognitive organization of nonhuman animals cannot accommodate itself to the kind of system constituted by human language, it will not be because the animals fail to score well enough on a specific suite of tests.

The communicative abilities of animals can be valuable to the cognitive scientist in several ways. To begin with, the behavior and abilities of animals are fascinating in their own right. As a simple illustration, Carolyn Ristau's study of the evasive displays of the piping plover makes it clear that this bird shows a level of apparently intentional behavior whose exact characterization in relation to the categories of human cognition presents intriguing challenges. Cheney and Seyfarth's studies of vervet monkey vocalizations (discussed in more detail in Chapter 7) have led to interesting conclusions about the limits and the possibilities of the animal's cognitive world, the individual's awareness of other animals, and much else.

We can learn about an animal's cognitive capacities not only by studying what it does when communicating with other animals, but also (potentially, at least) by utilizing this same system of communication as a tool to explore cognition directly. For example, if it were possible to teach an animal to respond meaningfully to "same-different" questions, the answers could be used to gather information about just what objects or phenomena in the world seem equivalent to (let us say) a parrot. This scheme may seem far-fetched, but we will see in Chapter 11 that approximately these results have been achieved with an African grey parrot named Alex. More ambitious hopes that animals can tell us the details of their lives and experiences appear for the present to be the stuff of science fiction, not science — or else of serious overinterpretation.

Finally, the very enterprise of asking how human language and animal communication differ does not tell us only about the ways in which the animals fall short of us. Focusing on what it is about humans that fits them to acquire and use languages tells us much about ourselves as well, because languages in the human sense are systems not known to exist in any other organisms.

How do we go about exploring the cognitive capacities of animals, especially their abilities in the domain of communication? Much as we would like to (and despite scientific conferences organized around the possibility), we cannot really talk about these topics with the animals themselves. We have to rely on our interpretations of their behavior under controlled conditions. But in making sense of what we observe about animal behavior, we face a serious problem known as the Clever Hans phenomenon.

The story of Clever Hans (Figure 3.1) has been told many times, perhaps in greatest detail by Oskar Pfungst. The horse's trainer, Wilhelm von Osten, maintained that the reason horses (and other animals) displayed less intellectual capacity than humans was primarily their lack of educational opportunity. He set out to rectify this omission with Hans, training him in a number of basic skills, including arithmetic (whole numbers, fractions, decimals), object identification, spoken and written German, days and dates, and standard German coinage. He then presented exhibitions of Hans's abilities, demonstrations in which the horse responded correctly to questions posed in German by tapping his hoof and shaking his head a number of times corresponding to the correct answer.

Figure 3.1 Clever Hans and Herr von Osten

Audiences of course were skeptical and attempted to demonstrate that Hans was actually receiving signals from von Osten (whose belief in the animal's abilities appears to have been quite sincere). This possibility seemed to be excluded, however, because Hans performed essentially as well when others posed questions to him in the absence of his trainer. In 1904 the director of the Berlin Psychological Institute, Carl Stumpf, established a commission to assess Hans's abilities. The thirteen members were selected so as to ensure that no deception would go undetected. Chosen to cover as many bases as possible, the commission included a circus manager, a teacher and a school administrator, two zoologists, a veterinarian, a physiologist, a psychologist and a politician. In spite of their skepticism, there seemed no sign of signals being passed (intentionally or not) between von Osten and his horse. Although Stumpf later insisted that the commission members were not at all "convinced that the horse had the power of rational thinking," they could say nothing more than that further investigation was warranted.

Oskar Pfungst, unwilling to let the matter rest, continued to test Hans. Eventually he established that the horse's abilities were indeed illusory. For one thing, if his questioner did not himself know the answer to the question posed, Hans performed much less well. He was also unable to answer when he could not see the person who asked the question, or someone else who knew what the question was.

Pfungst published his own report in December 1904. As it turned out, Hans would start tapping on the basis of one set of quite unconscious signals from his questioners and keep going until he got another signal to "stop." That is, he was not adding, making change, or understanding the question in any relevant sense. He was picking up on something totally different: unconscious involuntary movements of the head and body, small changes in facial expression, perhaps changes in his audience's heartbeat — completely unintended events that accompanied the listener's changing internal state as the "answer" to the question unfolded. Pfungst was even able to get Hans to start tapping his hoof by standing in front of him and making appropriate slight movements, without posing any question at all; the tapping stopped when Pfungst straightened his head slightly.

Clever Hans was clearly doing something rather more interesting than Nino (in the epigraph at the start of this chapter). No one had ever conspired with him about a set of signals: he worked them out on his own, without his trainer's even being aware of the external indicators Hans was detecting about his internal state. Certainly Hans did not need to know German, or arithmetic, or any of the other skills he was supposedly exhibiting in order to perform successfully. In the words of a newspaper story of the time, "The horse of Mr. von Osten has been educated by its master in the most round-about way, in accordance with a method suited for the development of human reasoning powers, hence in all good faith, to give correct responses by means of tapping with the foot. But what the horse really learned by this wearisome process was something quite different, something that was more in accord with his natural capacities, — he learned to discover by purely sensory aids which are so near the threshold they are imperceptible for us and even for the teacher, when he is expected to tap with his foot and when he is to come to rest."

The story of Clever Hans points out two morals that must be kept in mind when considering the interpretations we give to the behavior animals exhibit under even the best controlled conditions. Investigators may unwittingly give cues to their subjects, so that the observed behavior is actually determined by something quite different from what the experimenter is trying to explore. This prompting need not be intentional, and it can happen in a variety of ways, even those completely removed from the communicative modality on which the experiment is concentrating.

Furthermore, we cannot be sure that the animal's conception of what is going on in a given situation is anything like ours. Some of the early work done to explore the language abilities of chimpanzees employed artificial systems such as keyboards or plastic tokens as the communicative medium, rather than spoken words (or the manual signs that would later replace them in this research). In these experiments the animals learned to produce "utterances" that we can translate as "please—machine—give—Lana—M&Ms," each such utterance consisting of a specific sequence of symbol manipulations. Has Lana therefore come to control the structures of sentences like the English ones we give as "translations"? No. Lana has merely learned that the signal that elicits M&Ms from the machine has a complex form. There is no reason (at least not without a lot more research) to say that the key or plastic token which *we* translate as "please" has any signification for Lana or the other animals in these experiments such as what the English word means to us: she is merely doing something that gets her M&Ms.

It is possible to train pigeons to peck a sequence of several different levers in a particular order to obtain a reward, a behavior that we would surely be misguided to interpret as the production of structured "sentences." The temptation to impose such interpretations on animal behavior is sometimes irresistible: classic parrot utterances ("Polly want a cracker") seem quite definitive in their interpretation to us, but there is almost never reason to believe that they have any such interpretation for the bird. The case of Alex is quite different: Alex does mean what he says, at least in large part, but he is a special case.

In light of these difficulties, it is generally assumed that observations of an animal's behavior should be interpreted in terms of a very conservative principle known as Morgan's canon: "In no case may we interpret an action as the outcome of the exercise of a higher psychical faculty, if it can be interpreted as the outcome of the exercise of one which stands lower in the psychical scale."

There are obvious problems with this formulation. What is a "faculty"? What makes one faculty "higher" or "lower" than another? Anyway, why should we believe in such a restriction? But by and large, Morgan's canon simply corresponds to a moderately conservative notion of what constitutes a scientifically warranted conclusion.

Suppose you are trying to show that animals have some particular ability, and they emit some behavior that is consistent with their having that ability. If their behavior is also consistent with some other assumption, you have not shown that your preferred interpretation is correct until you have excluded the alternative. Clever Hans's behavior was *consistent* with his having learned German, arithmetic, and so forth; but it was also interpretable in terms of his simply having learned to pick up subtle start-stop signals. The more elaborate interpretation has not been established until the simpler one is ruled out. And since it is fairly easy to show that a horse can indeed learn to start and stop tapping his hoof on the basis of signals, it is more conservative to interpret what Clever Hans was doing in that way. Of course, the more elaborate story is still logically possible, too; but we have not shown it to be correct unless and until we can rule out the simpler one.

We do not, perhaps, need anything as grand as Morgan's canon to decide what to say about this case, but other examples are subtler. I have already mentioned another classic of the experimental and philosophical literature on the cognitive capacities of animals: the behavior of some birds, including the killdeer and the piping plover, who seemingly pretend to be hurt in order to distract predators who might endanger their nestlings. During the nesting season, when such danger approaches, the mother bird attracts the intruder's attention with a "broken-wing" display (Figure 3.2) while moving away from the nest site. Seeing this apparently wounded animal as easy prey, the predator follows the mother until both are safely out of range of the nest, at which point she reveals her true state of fitness by flying off.

Such behavior seems to wear its interpretation on its sleeve: surely the bird is actively feigning injury with the intention of misleading the predator, drawing it away from the helpless chicks. But do the facts really lead us to that conclusion? Think of what is involved in actively deceiving someone else. First you have to imagine how the other individual interprets the world as it is. Then you have to work out that *if* the other were presented with certain (apparent) facts, that interpretation would be altered (in a way that you personally would prefer). And then you have to implement some course of action that will seem to present the relevant "facts" to the other for this purpose. Quite a lot of cognitive computation, after all.

Ristau examined in detail the plausibility of this interpretation of the piping plover's deceptive display. There is little reason to doubt that poten-

Figure 3.2 The piping plover's broken-wing display

tial predators are effectively deceived by the bird's broken-wing display and lured into following the mother away from the vulnerable nestlings. But does this necessarily entail that the plover actually intends to mislead? Although undeniably an effective trick, the behavior in question is nonetheless a somewhat isolated element in the bird's behavioral repertoire. We have no evidence that the plover deceives other animals (or other plovers) about other things in other ways.

The simplest possible interpretation of the bird's actions would be that the broken-wing behavior is no more than a reflexive and completely automatic response to danger. Ristau shows that cannot be correct, however. When the predator does not follow the mother, she comes back toward it, repeats the display, and moves away again, until finally she attracts its attention.

Further, other intruders (animals such as cows that do not eat the eggs but might step on them) do not provoke this behavior—although if they get too close, the bird may fly up in their face to scare them away. The bird monitors the intruders, and if they do not get drawn away, she goes back and tries harder. It seems to be exactly those intruders that present a real danger to her chicks that she is trying to lead away from the nest. Thus, her actions cannot be the result of a simple reflex to twist her wing and move away when others approach the nest.

Apparently the bird has the intention to draw intruders away from her nest. But that does not determine the cognitive mechanisms that underlie the broken-wing display. Indeed, that behavior has variants. Sometimes the bird just makes a lot of noise and flaps her wings while heading away from the nest, attracting the intruder's attention. There seems no reason to interpret the broken-wing behavior as specifically and intentionally deceptive (in the rather elaborate sense laid out above). It is just another way to draw the enemy away from the nest.

The difference in interpretation is subtle, but crucially important in assessing the kind of cognitive processes we want to attribute to the plover. As with Clever Hans, we have not excluded the possibility that the bird has an intention that involves an understanding of the predator's interpretation of the situation—but we have not shown that, either. It is simpler to assume the bird adopts this behavior as an effective diversion from her eggs. Without any understanding of why it works as it does, we must adopt that interpretation until it can be shown that something more elaborate is at work.

What is at stake here, in the jargon of the field, is whether we need to assume that birds have a "theory of mind," by which they attribute interpretations and attitudes to other beings (and sometimes seek to influence their behavior by affecting the content of their minds). We can fairly easily convince ourselves by simple introspection that *we* work that way, but it is notoriously hard to explore this issue in others.

Experimental evidence suggests that an understanding of the minds of others emerges early in human infancy. Autistic children, on one view, differ from normal children primarily in lacking this kind of understanding of others. Daniel Povinelli has argued that chimpanzees fail at tasks in this area that young children easily pass. If he is correct, we should probably not attribute a theory of mind to these primates. Before we can adopt a view of behavior as genuinely deceptive, we need to show that the animal in question has a conception of other animals as having minds whose content can potentially be influenced. Such evidence is not at all easy to come by, and probably is presently lacking for all species other than our own.

Charles Munn describes two species of birds in the Amazon that provide a similar example. In mixed-species flocks the white-winged shrike-tanager and bluish-slate antshrike serve as "lookouts" while others feed, because they are generally the first to give alarm calls when dangerous hawks approach. The other birds respond to these calls by freezing, looking up, or taking cover. Sometimes, though, when there is competition for the insects all the birds feed on, the sentinel birds give alarm calls when no danger at all is present. The others disperse, leaving the crafty watchmen a clear path to the food.

Should we interpret this activity as the result of an *intention* on the bird's part to deceive his non-conspecific companions? Probably not. The more conservative view is that the bird gives the alarm because doing so (occasionally—it would not do to overuse this strategy) provides unencumbered access to food. Accordingly, there is no reason to believe that the calling bird has any particular sense that the other birds "believe" anything, and thus that it might be possible to "mislead" them.

In Ristau's piping plover study, the initial observations were consistent with the bird's acting in the way it does as a reflex; or with her having a fixed behavioral routine; or with her having an intention to lead the intruder away; or, finally, with her aiming to deceive the intruder. Ristau's subsequent observations showed that the behavior has a flexibility that is inconsistent with reflexes or fixed repertoires, allowing us to conclude that some sort of intention is involved, and that it entails leading the intruder away. There is, however, no evidence to require the conclusion that deception is a part of this intention.

Each step along this path involves an application of Morgan's canon, at least implicitly. Sometimes the result is that yes, some interesting structure is necessarily entailed by the facts surrounding the behavior under study. At other points we must conclude that no data support going further in our interpretation. That does not in itself constitute evidence that some additional structure is *not* present: it is perfectly possible, at least logically, that plovers are really a kind of robot, controlled from inside by devious little green individuals from Mars, who do indeed intend to toy with the minds of the other animals their hosts encounter. But in the present state of our knowledge, we cannot regard *any* interpretation that involved intentional deception as scientifically justified.

As we read the literature interpreting animal communication and explore the attempts that have been made to teach language to animals, we must bear these methodological issues in mind. We want to avoid inflated claims and excessive romanticism; but we also want to give the animals their due, not treat them as mindless simply because they are not human. Questions of the cognitive complexity of other animals are at least in principle a matter for serious scientific investigation, and the results that emanate from such study are not at all obvious in advance.

However, in developing a serious science of animal cognition we must resist the temptation to limit our questions to those based on human capacities. In the domain of communication, if we ask "Can nonhuman animals learn a human language?" the answer is likely to be (as I argue in the rest of this book) that they cannot. If we let the matter rest there, however, we will undoubtedly be ignoring the vast range of other possible questions about animals to which the answers might well be quite different—and far more interesting.

Human Language: A First Look

We need to be conservative when interpreting animal behavior if we want our conclusions to have some kind of scientific status. Behavior that seems on the face of it to be quite complex often is somewhat simpler, at least as far as serious inquiry can determine. I would like now to examine the opposite possibility: behavior that seems on the face of it so simple and natural as to call for very little in the way of specialized cognitive underpinnings sometimes proves on closer inspection to be vastly more complicated than it appears.

The behavior in question is what we do when we produce and understand sentences of our native language. What could be more effortless? While we may spend considerable time and attention in deciding what to say, no comparable overhead is apparently involved in how we say it. Similarly, we may have trouble in understanding the ideas expressed by another person, but (except for specific, unfamiliar words, or speakers with an incomplete command of our language) that seems to be a matter of the ideas and not the communication system.

A human language is an extremely elaborate and detailed system, so highly structured and yet so comparatively uniform across its speakers that we can imagine no realistic alternative to the assumption that its system is deeply determined by our biology. In later chapters I explore specific aspects of that complexity, as it plays out in sound structure, sentence structure, and the path by which knowledge of our native language arises. For now, the point is simply that there is a great deal to be explained about how our knowledge of language arises and is put to use.

The Nature of Natural Language

How does linguistic communication work? How does it accomplish its goals? Suppose Mary says to John:

(1) Animals have many fascinating abilities, which we should not confuse with our own.

The behavior Mary exhibits here is a series of coordinated movements of her vocal organs, movements that result in the production of sound. The range of messages Mary might convey by such movements is vast, surely infinite. The richness of the message space should be fairly obvious: small variations in the form of the sentence lead to quite different messages, in ways that seem to have no limits.

John, for his part, can be pretty sure that Mary has a fairly structured thought, and that she really did intend to convey the thought represented by (1). In some situations, though, her intention might be related only indirectly to the literal content of her utterance. A sentence such as "How do you do?" is not usually uttered as an inquiry about the manner in which the listener might perform some action, but rather as an acknowledgment of an introduction. Complex relations between message and content may involve irony, as in "Sure, you hacked into the NYT Website. And I'm the Easter bunny." Or even "I love you."

The meaning of (1) is certainly not limited to reactions to the here-and-now. The first part of the sentence describes a situation as true *in general,* not just in the temporal neighborhood of Mary's utterance. The second part refers to her attitude toward something that she hopes will not occur. Messages of either sort (expressions with general, not particular, relevance; or references to nonexistent but possible events) are not found, as far as we know, anywhere outside of human language.

How does this sentence convey the meaning it does? Central is the fact that (1) is not just an isolated signal, but forms part of a comprehensive *system* of language, a language of which both Mary and John are speakers. To be a speaker of a language, a person has to have a certain sort of *knowledge,* by virtue of which it is possible to understand the message conveyed by particular sentences. If John knows only Georgian, for instance, and not English, Mary's utterance of (1) will be quite ineffective in conveying her meaning.

What does John need to know to qualify as knowing English? A part of the answer is that John must know the meanings of English words. For many, knowledge of a language is virtually equated to knowledge of its words and their meanings; but surely there is more to it than that. For instance, the same collection of words arranged in alternative ways may differ considerably in whether or not they constitute a possible sentence in the language—compare (2a) and (2b)—or in what the sentence as whole means, as in (2c) and (2d).

(2) a. This morning Fred is the hall monitor.
    b. Monitor hall the is Fred morning this.
    c. Mary believed John's mother to be the murderer.
    d. The murderer believed Mary to be John's mother.

Apparently John's knowledge of English must include more than the words of the language and their meaning. He must also know what is systematic about their arrangement into larger meaningful units such as sentences. And what he knows about this process is (at least in part) a reflection of the fact that he knows *English.* If he knew Georgian, he would know some partly different things. Georgian has different words that express (largely) similar meanings, and different principles of arrangement; knowledge of the one system does not in any way follow from knowledge of the other (as you can convince yourself the next time you are in Tbilisi and need to ask directions).

On the basis of even this limited understanding of the nature of language, we can begin to lay out some of the aspects we would have to explore in order to ask the question whether some nonhuman animal is capable of language. Obviously, to determine that, we have to understand what language is. And that involves a number of subquestions:

- What is the system of knowledge a speaker has? What kinds of knowledge are involved, how are these organized, to what extent is this knowledge specific to individual languages, and to what extent is it common across all languages?
- How does this knowledge arise in the mind? Do we learn it? In what sense of "learn"? Or does it just grow there? What is the interaction between our biologically determined organization as human beings and our experience at various crucial points in life that yields the knowledge we have of our language?
- How is this knowledge put to use? How do we utilize our broad knowl-

edge of our language in understanding others and in producing sentences ourselves?

- What physical mechanism is the material basis for this knowledge and its use? Presumably, some aspect of our brains, but which? To what extent is the tissue that supports knowledge of language specialized for this function?

These are all enormous questions, and I cannot hope to answer them all in this book. My purpose here is more limited: we need only enough detail about human language to support the broader comparison, and contrast it with the capacities of nonhuman animals.

As my story unfolds, I will return repeatedly to a thematic proposition of which a great many linguists have become convinced, especially since the 1950s or so. This is the notion that the form taken by our knowledge of language is determined to a substantial extent by our biological nature as humans. We come into the world "preprogrammed," as it were, to develop a particular kind of knowledge (that is, to learn a language), on the basis of a particular kind of experience (hearing what goes on around us in the speech of others), at a particular stage in our development.

We develop a knowledge of language that closely resembles that of our surrounding community not on the basis of a decision—or even of a conscious motivation—to do so, but rather as a natural aspect of maturation. In this sense, we do not really *learn* our native language; rather, we *grow* it, in the way we grow hair on various parts of our bodies at particular stages in our physical development. Of course, we need exposure to the speech of others in order to develop a particular language, but astonishingly little is actually necessary to trigger the growth of linguistic knowledge.

The growth of language is thus similar in some respects to that of other cognitive faculties, such as vision. It involves an interplay of our biological nature and specific kinds of "triggering" experience, as we discussed in Chapter 2. When linguists assert that "language is innate," this is what we mean—not that the knowledge of, say, English or Georgian is already present in the womb. When properly understood, this proposition really ought not to be so controversial. Who would seriously question the claim that the way our visual system develops is innately determined? Like much concerning language, though, the topic tends to arouse the passions of many.

We can find abundant reasons to believe the apparently inflammatory claim that language is a basic part of human nature, determined like other aspects of human nature by human biology. Many of these reasons are not particularly inaccessible. Steven Pinker's highly readable book *The Language Instinct* provides an excellent survey.

One argument that seems to connect language firmly with human nature is the *universality* of language: every community ever encountered has one. Groups of people isolated from other human contact may develop in unique ways, but those do not seem to include the possibility of giving up language, or of failing to develop it, no matter how deeply the isolation is rooted in human prehistory. To be human is to have a language — at least one — except in cases of severe individual pathology.

Actually, in much of the world the natural condition of humankind is multilingualism. Many individual groups come in contact with one another regularly enough to warrant general cross-fluency in multiple languages. This contact does not generally affect the individuality of the various languages involved, each of which retains vitality within its own sphere.

One might well say that every community has a language because the original human society from which the species descends had one, and it has never been given up. That statement is inadequate, however, because groups do occasionally arise in which connection with the ancestral language has been severed. (I have in mind groups of congenitally deaf individuals, for whom access to the spoken language of a surrounding community may be completely cut off.)

In many cases, hearing-impaired people *do* have access to a language, one that operates in the visual-spatial modality, rather than auditorily, such as ASL. Although Hockett's proposed set of design features for language would exclude such a system from consideration as "language," systems like ASL do indeed have all the characteristic structural properties we associate with spoken languages, apart from modality. And it is evident from a number of cases that signed languages develop spontaneously and remarkably quickly in communities cut off from spoken language.

Of course this does not mean that an isolated deaf individual, or even a small number of such people, will immediately produce a fully structured manual language ex nihilo. Under moderately favorable conditions, though, the time involved can be remarkably short. The best-documented case is the emergence of Nicaraguan Sign Language within little more than a decade. The work of Susan Goldin-Meadow and her colleagues on the spontaneous development of signing in hearing-impaired children with no language input has shown how strong the urge to develop language can be, even in the absence of appropriate experience.

The universality of human linguistic accomplishment presents an interesting puzzle, in light of the apparent naturalness of signed languages. Why has science never identified a community in which the local language is a manual one, even though the people show no greater incidence of hearing impairment than does humanity at large? If signing is as viable a channel for language as speaking, why is it never the vehicle of choice where both options are equally open?

Perhaps this is where the historical argument has a bearing. It is hard to imagine circumstances in which an isolated group of people would lose access to the language of an earlier community from which they were descended. Deafness can certainly block transmission, but if some original group of humans had a spoken language, it seems likely that it would persist through tradition alone. It would change, of course, as all languages do over time, but each generation would continue a language close to that of their parents. Perhaps this explanation is sufficient—but it seems to fall short, somehow. We may need to recognize a bias toward the spoken modality in language unless it is excluded by physical handicap.

A number of other observations have become part of a standard menu arguing for the deep roots of language in human nature. Consider, for example, the development of those (fortunately rare) communities in which individuals from various backgrounds are forced together in a situation where there is no common language to facilitate communication.

What sometimes happens under these circumstances is that a "pidgin" develops. The system has little or no linguistic structure beyond that of a collection of shared words sufficient for the minimal needs of basic communication. Where such communities persist long enough for offspring to be born into them, the children quickly develop a fully structured language (a "creole")—even though the structural principles of this language were not present in the input that made up their linguistic experience! This observation is important because the principles of linguistic structure that emerge in the development of a creole from a pidgin are not grounded in the learners' experience: they must come from somewhere else. The "somewhere else" is presumably the biologically determined cognitive organization of the learners themselves.

It is often suggested that even if language is based on the nature of human cognition, no special principles unique to language are required. All that is necessary is available as particular instances of general cognitive abilities, such as learning, memory, problem solving, and other functions that are not domain specific. This proposal, however, seems implausible in light of the empirical evidence. Cases of pathological cognitive impairment, in particular, show a distinct separation between linguistic and other, general functions.

The literature on these matters is too vast to present here. The bottom line is that language and general cognitive ability show a double dissociation. Severe linguistic impairment may occur in the presence of otherwise normal overall intelligence and cognitive skill. The reverse is true as well, in that language can be quite normal in individuals whose general intellectual functioning is greatly impaired.

Apart from cases of aphasia resulting from specific brain injury (which may well reduce language function severely while leaving most other cognitive functions intact), an interesting example of the first dissociation above is found in cases of so-called Specific Language Impairment (SLI). This term refers to a particular pattern of damaged language skills that is apparently traceable to an inherited genetic defect. The members of a family who are affected by SLI differ considerably from their relatives in linguistic abilities, but not in other cognitive areas.

The opposite pattern, in which language ability remains intact even in the presence of serious impairment, is found in children with Williams syndrome. Although these children do not do well on tests of general cognitive ability, they produce language that is, in structural terms, entirely normal — even exuberantly complicated.

A similar instance is provided by Christopher, a "linguistic savant" whose abilities have been studied in considerable detail by Neil Smith and his colleagues. Christopher functions very poorly in general; he scores between 46 and 72 on IQ tests and is unable to tie his shoes without aid, for example. But he excels in one area: the learning of languages. He has learned French, Spanish, Greek, Berber, and more than a dozen other languages quickly and quite well. Most recently, he has learned a considerable amount of British Sign Language (a manual language unrelated to American Sign Language), although his lack of physical coordination presents special difficulties.

Taken together with the evidence for language as based on our biologically determined nature, the further fact that language ability and general intelligence are so poorly correlated gives us reason to believe that the relevant part of that nature is specific to the domain of language. It should not come as a great surprise that our biology is specialized in this way. Alligators, for example, have a special set of sensors on the skin of their faces that respond sensitively to the slightest disturbance of the surface of a body of water in which the alligator is mostly immersed. Catching (and eating) the sources of such disturbances is one of the alligator's most useful (and characteristic) skills, and there is no serious doubt that this skill results from specific, inherited aspects of the animal's biology. Similarly, the use of language to communicate is one of humankind's most useful and characteristic skills, for which a comparable account is no less plausible.

The Complexity of Natural Language

I promised to provide evidence that linguistic abilities are subtler and rather more complex than they appear, but thus far my discussion has focused on reasons to believe those abilities are part of our biologically determined nature. Of course, the more deeply rooted in species-specific properties language is, the more likely it will show unusual quirks and hidden structure. It is time now to address some of the hidden complications that pass unnoticed in our everyday use of language.

Consider, for instance, a fairly simple English sentence:

(3) Mary expected to see her.

What do we know about the meaning of this sentence? We recognize the individual words and know their meanings. We also understand the structure: *Mary* is the subject of the verb *expect,* the pronoun *her* is the object of *see,* what Mary expects is *to see her. Mary* is not only the one who expects something, but also the one who is to do the seeing.

Let us focus on the pronoun *her* in this sentence. Pronouns, of course, refer to some individual, and the same pronoun can refer to different individuals in different sentences. What do we know about the referent here of *her?* As a function of the lexical meaning of the word, we know it refers to a female individual, but that is not all. While the context in which (3) is uttered might provide some information, *her* in this sentence could in principle refer to *any* female individual—with one exception: *her* could not refer to the same person as *Mary.* This obviously has nothing to do with the basic meaning of *her*. In a sentence that differs minimally, say *John expected to see her*, the same word in the same grammatical position can perfectly well refer to Mary.

Now consider:

(4) I wonder who it was that Mary expected to see her.

Sentence (4) appears to contain (3) as a proper subpart, so we might expect the parts of the latter to behave in the same way, but that is not what we find. In (4) *her* can perfectly well refer to Mary (though it need not — it could also refer to some other female).

This somewhat surprising contrast between the meaning of *her* in (3) and (4) is one about which speakers of English can agree, regardless of their origins, education, or literary inclinations. It is natural to ask, therefore, where knowledge of this sort comes from. Has someone taught every speaker of English this rather subtle fact about the use of pronouns? Surely no one has been taught how to use *her* in these specific sentences, because the same statements are true even when we have never heard the sentence before (consider these sentences with, say, the verb *repudiate* substituted for *see*, where the interpretation of *her* remains unchanged). Almost as surely, no one has been taught the general principle (apart, perhaps, from the children of a few eccentric linguists), since these facts had not been noticed as a systematic part of language before about 1965.

Perhaps there is no great cosmic principle involved. Perhaps speakers simply understand these sentences "by analogy with" other sentences they already know how to interpret. But the notion of analogy is something of a tar baby. Far from rescuing us from a responsibility to uncover the precise principles of English grammar, it simply gets us deeper into trouble. For instance, consider the pair of sentences in (5):

(5) a. My cat can speak Georgian.
    b. My cat can speak.

What does the second of these sentences mean? Apparently, it means that my cat can say some arbitrary thing. It looks as if, in a sentence whose verb is one such as *speak* that takes an object, we can omit mention of that object with the result that it is interpreted as something arbitrary (compare *Pierre is eating foie gras* with *Pierre is eating* and the like).

If that is the case, what about the sentences in (6)?

(6) a. My cat is too sensitive to tolerate disorder.

    b. My cat is too sensitive to tolerate.

What does the second of these mean? How did you know? And why isn't the "arbitrary object" meaning available? We ought to be able to interpret this pair "by analogy with" sentence pairs like (5), but we cannot.

    The point is that notions like analogy do not really get us very far. Why are some analogies valid while others are not? The valid analogies are the rules or principles that characterize our knowledge of our language, and those are quite specific. The difference between those that work and those that do not does not in itself follow from the overall nature of "analogy." We do not get the correct principles for free as instances of some general principle(s) of inference or logic. And almost certainly, no one teaches them to us.

    Now consider some more complex sentences:

(7) a. Monica is too embarrassed to tell anyone to talk to Bill.

    b. Monica is too smart to visit anyone who talked to Bill.

The interpretations of these sentences are reasonably straightforward and not seriously in doubt: Monica is the one who is not going to do the telling or the visiting, whereas Bill is the one who is to be (or was) talked to. That is, *Monica* is understood as the subject of the *to*-phrase, and *Bill* is the object of *talk to*. Now look at the sentences in (8):

(8) a. Monica is too embarrassing to tell anyone to talk to.

    b. *Monica is too smart to visit anyone who talked to.

In the first of these, *Monica* is understood as the object of *talk to,* and some arbitrary person is the understood subject who is to do the telling. We might then expect the second sentence in (8) to be analogous to the first, with Monica as the object of *talked to* and an arbitrary person doing the visiting. But in fact, this sentence is garbage ("word salad"—or to use the more sober technical terminology of linguistics, *ungrammatical*): that is what the * indicates. In later discussions of the syntax of natural languages, we will see a lot of these *s, because it is often as important to study the things that are *not* sentences as the things that are.

    The ill-formedness of (8b) raises two problems. First, why isn't this sentence understood on the basis of one of the possible analogies with others?

Its individual pieces all have parallels in sentences of impeccable character, but somehow that fact does not provide us with a basis for understanding this sentence.

Second, how can it be that this sentence is excluded? We know what the words mean. We have a basis for assigning interpretations to phrases at various levels. Yet the sentence is simply bad. The problem is not that it means something other than what it should: it is just ill formed altogether. In the other cases we considered, we might have thought there are several possible analogies and that one is "stronger" (in some sense, to be made clear) than the others, and thus that one "wins," perhaps obscuring the other possibilities. In (8b) it appears that the system of English that forms part of our knowledge simply does not assign an interpretation to the sentence.

The "sentence" in (8b) is an ungrammatical string of English words. The very possibility of this situation is related to another point. We can recognize some sequences of words as conforming to the grammar of the language even though they fail to correspond to anything meaningful. The classic example is the pair of strings in (9) made famous in Noam Chomsky's early work—so famous, in fact, that (9a) is Chomsky's only entry in recent editions of *Bartlett's Familiar Quotations*.

(9) a. Colorless green ideas sleep furiously.

    b. *Furiously sleep ideas green colorless.

Despite both being incoherent, these strings have quite different status. (9a) conforms to the grammar of English, although it does not mean anything sensible (despite the attempts of poets and others to construct discourses in which it could appear); (9b), just as meaningless, is word salad. The difference in status is readily apparent to a native speaker of the language, and makes it clear that grammaticality is not the same as meaningfulness.

Indeed, this difference goes both ways. We can often make a surprising amount of sense out of strings of words that are not, in fact, grammatical in our language. Some examples of this in Japanese advertising copy, a famously rich source, are given in (10).

(10) a. (on an item of clothing:) Our clothes makes healthy and sexy impression to us. It transforms yourself completely, and giving you happy times.

   b. (on a package of drinking straws:) Let's try homeparty fashionably and have a joyful chat with nice fellow. Fujinami's straw will produce you young party happily and exceedingly.

   c. (in a Tokyo hotel:) It is forbidden to steal hotel towels please. If your are not a person to do such thing is please not to read notis.

Here we can make sense out of the sentences even though they are not put together in such a way as to be grammatical.

A person who knows English can assess strings of English words as, first, grammatical and/or, second, meaningful, where the two notions are independent of each other. This possibility has proven crucial to an understanding of precisely what it means to "know" a language. Until fairly recently, linguists concentrated on descriptions of the set of words, sentences, and the like that are part of a given language. We now see, however, that our knowledge covers not only what is in our language, but also what combinations of English words are *not* part of our language. Data about ungrammaticality in this sense are often as significant as data about grammatical sentences.

The knowledge we have as speakers of a language is not merely a collection of words, each a learned association between sounds we can produce (or manual gestures in a signed language) and something we (or another) might mean by them. Instead, it is a rich system of detailed and specific regularities. We often refer to these regularities as *rules*, but the sense of "rule" here is not that of the mythical English teacher Miss Fidditch's injunction never to split an infinitive. Nor are we concerned with rules in the sense of the rules of a game, which can be changed quite freely because they are not based on any nonarbitrary set of underlying principles. (Some might claim, however, that the American League violated some basic principle of nature in introducing the designated hitter rule in baseball.)

These regularities are specific to language, not simply extensions of our general cognitive abilities. Some are undoubtedly learned from early experience, at least in some sense, because different languages can have different principles of word order, different patterns of the relation between grammatical function and word form, and much else.

Other principles, such as those for the interpretation of pronouns in the discussion of examples (3) and (4), turn out to be effectively universal. Once we learn that a word of our language (for example, *her* in English) is a member of a particular general category (such as pronoun), much of its grammar follows from that fact alone without our needing to have the details pointed out explicitly—surely a virtue, in that our language models are quite unable to provide such assistance!

Further, much of the information involved in exploring our knowledge of language comes from the study of *un*grammatical sentences. Obviously, at least some of the sentences a child hears while learning a language are ungrammatical. These include hesitations, false starts, cases where the speaker forgets how the sentence began before finishing it, and so forth. Precisely how much literally ungrammatical data may be present in the input to the child is a matter of controversy, but if there are even a few such sentences, with no indication of just *which* sentences these are, my general point is sufficiently made.

The mature grammar that the child constructs on the basis of a complex mixture of sentences of varying grammaticality nonetheless allows the adult to make judgments about these sentences. What is remarkable is the fact that grammatical and ungrammatical sentences are inextricably mixed in the input, with no indication of the status of any one of them. Even without such "negative evidence," children come to know the difference.

As a result, much of what we know as speakers of, say, English literally cannot have been taught to us. The only plausible suggestion seems to be that these aspects of our knowledge are consequences of the way humans are structured: no real alternative exists. These principles are not "learned" in the way words are. But if at least some major parts of our knowledge of language follow in this way from our biology, it stands to reason that a creature with a different biology would have access to somewhat different kinds of knowledge—and not necessarily to that which arises quite naturally in us.

In this chapter we have dealt with two complementary aspects of the study of complexity in cognitive systems. On the one hand, we saw that apparently complicated behavior, which seems to demand interpretation in terms of elaborately structured knowledge and reasoning, is often susceptible to a much simpler explanation. If we want to tell an exciting story, we may opt for one based on rich mental structure. If we want to establish a scientific conclusion, however, we have to force ourselves to stop at the most direct and uncomplicated account possible. That is not because we have shown such a story to be the truth, but rather because nothing stronger can be justified by the available facts.

Still, this striving for scientific parsimony must not blind us to the fact that apparently simple behavior and cognitive processes may actually conceal great complexity that we can reveal only by detailed exploration. In particular, our ability as speakers of a natural language seems simple and effortless, based mostly on our having learned a collection of associations between sound and meaning. Looking only a little deeper, however, we see that a vast realm of intricately structured knowledge underlies our production and understanding of utterances in our language.

In later chapters, I explore some of the components of this knowledge, and some of what we know about how it arises in the organism. This focus on human language has considerable interest in its own right, but it is essential to our task of evaluating the cognitive abilities of nonhuman animals in the realm of communication. Unless we understand in detail exactly what it would be for an animal to control the system of a human language, we cannot seriously evaluate behavior that seems similar to ours when we use such a system.

4

The Dance "Language" of Honeybees

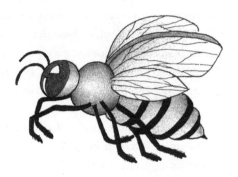

For my part, I cannot truthfully say that I ever got into real, personal conversational contact with the insect world. But that John Dolittle did, there can be no doubt whatever. This I have proof of from things that happened. You cannot make a wasp stand up on its front legs and wave its other feet unless you know enough wasp language to make him do so. And that—and a great deal more—I have seen the Doctor accomplish.

Of course it was never quite the free and easy exchange of ideas that his talking with the larger animals had come to be. But then insects' ideas are different; and consequently their languages for conveying those ideas are different.

—*Doctor Dolittle's Garden*

Bees have a surprisingly rich history of posing important questions for science. For example, in 1934 the entomologist Antoine Magnan wrote that his laboratory assistant, André Sainte-Laguë, had applied some standard assumptions of fixed-wing aerodynamics to the bumblebee and concluded that flight was not possible for such a system. Thereby was born the urban legend that "according to the theory of aerodynamics, as may be readily demonstrated through wind tunnel experiments, the bumblebee is unable to fly. This is because the size, weight, and shape of his body in relation to the total wingspread make flying impossible. But the bumblebee, being ignorant of those scientific truths, goes ahead and flies anyway—and makes a little honey every day."

Although the bee's behavior under such apparently adverse conditions has thus become the basis of an inspirational poster (from which the quote is taken), the truth is more prosaic: when the calculations are made correctly, the relevant aerodynamic conditions are simply not those that Saint-Laguë assumed. All the same, the investigation of this problem has been productive for understanding both bumblebee physiology and the physics of flight.

Similarly, one of the staples of the study of communication for many years has been the behavior by which honeybees (*Apis mellifera*) forage. Under appropriate conditions, the bees perform "dances." A small set of parameters of the dance rather precisely characterize the location of pollen, nectar, or (during swarming) potential nesting sites. These parameters correlate closely with the distance, direction, and rough quality of a recently visited food source or nesting site. Observers have found that other bees present at these performances turn up at the relevant location fairly soon thereafter, leading us to the conclusion that the dancer communicates this information by means of a highly structured code.

It seems entirely natural therefore to speak of the "dance language" of bees. Even for sophisticated observers, "the dance communication system is called a language because it satisfies all the intuitive criteria that have been posited for a true language." But surely we should not stop at such "intuitive criteria" any more in the case of language than we would in aerodynamics or entomology, given that scientific frameworks exist within which the relevant questions can be posed in all of these domains.

There are a number of logical leaps in the "bee language" story, and filling in the gaps will prove instructive with respect to both bees and language. To begin with, one bee dancing and another later arriving at the food source does not by itself show that the linguistic intentions of the first bee are the cause of the productive foraging of the second one. The first bee's behavior probably *does* communicate the location to the second, but that is not self-evident. The attempt to prove it conclusively has led to ingenious experiments that bring out much more as well. And when we ask what is the relationship between the properties of this apian communication system and those of human language, the distinctiveness of the latter stands out in bolder relief than might otherwise be the case.

The History of the Problem

Large portions of the description of the honeybee's dance are uncontroversial. First, it is exclusively the (female) worker bees, as opposed to the queen(s) and the (male) drones, who do the work of the hive: building honeycomb, nourishing and nursing larvae, defending the hive against intruders, getting rid of the bodies of dead bees, and other tasks. The duty of greatest interest to us is that of foraging for sources of pollen and nectar. When a forager bee finds an attractive food source, she returns to the hive and may dance around in an agitated fashion. Other bees gather around her. Shortly thereafter, a number of these other bees may turn up at the same food source and gather nectar or pollen there for as long as it seems productive to do so. Worker bees that have been recruited to a food source in this way may also dance when they return, and subsequent flights of recruits may continue to visit the source.

This much has been known for a long time. Aristotle, who was interested in bees, has a passage in his *Historia Animalium* that is rendered in English as follows by Karl von Frisch:

> On each trip the bee does not fly from a flower of one kind to a flower of another, but flies from one violet, say, to another violet, and never meddles with another flower until it has got back to the hive; on reaching the hive they throw off their load, and each bee [on her return] is followed by three or four companions. What it is that they gather is hard to see, and how they do it has not yet been observed.

The interpretive crux of this passage is the word *parakolouthoûsin*, "they follow (closely)." One interpretation might be that when the original forager, after dancing, returns to the field, she is *followed by* others. The dance itself might be simply a way of attracting the attention of other foragers, who would then need to be led to the goal. More plausibly, the word means that during the dance, several other bees directly follow the dancer's movements. It still is necessary to elucidate the relationship between the dance and the subsequent behavior of the bees, but the connection might be more complicated than simply drawing their attention to the successful foraging colleague.

So far, nothing very dramatic has emerged. It seems likely that Aristotle saw the dance as a way for the first bee to attract the attention of the others, who might then follow her back to the food source. At least as early as the eighteenth century, though, observers noted that the bees' dance on returning to the hive was not simply a matter of running around in disorganized fashion, but that it had a kind of pattern. Accounts by a variety of eighteenth- and nineteenth-century scholars posed the possibility that the other workers could determine from this pattern some aspects of the food source involved, and that the bees' dance was a kind of "language" by which the forager described the location of the food. This was immediately seen as a critical issue. Sir John Lubbock pointed out:

> Everyone knows that if an ant or a bee in the course of her rambles has found a supply of food, a number of others will soon make their way to the store. This, however, does not necessarily imply any power of describing localities. A very simple sign would suffice, and very little intelligence is implied, if the other ants merely accompany their friend to the treasure which she has discovered. On the other hand, if the ant or bee can describe the locality, and send her friends to the food, the case is very different.

Lubbock himself failed to find evidence for any special communication among bees beyond what would be expected on the basis of such "a very simple sign."

Maurice Maeterlinck was the first to consider the bee language question in terms of experimental evidence. He concluded from his studies that the other bees cannot find a food source simply on the basis of what they derive from the dance. They have to follow the successful forager back, as implied in Aristotle's description, or else they go out and look for food of the sort found by the first bee. There is no difficulty in identifying the relevant flower type, of course, because the first bee comes back redolent of some particular kind of nectar or pollen (recall the constancy of their attention in this regard, as noted by Aristotle).

Bruce Lineburg, who also considered the possibility of bee language, came to much the same conclusion. He argued that the mechanism at work involved the recruits following subtle trails of odor in the air, rather than their being instructed communicatively by the dance. On this view, the role of the first forager bee's dance is to excite attention and alert others to the existence of a food source.

This possibility is not at all out of the question. Bees are extremely sensitive to some odors, and odor alone can direct them over fairly long distances. For instance, they secrete a specific pheromone from the Nasinov's gland in their tail. Under conditions of high winds and heavy weather, a number of foragers from a hive may be stranded in the field, unable to buck the wind to return home. When this happens, other bees who have managed to make it back may produce the Nasinov's gland pheromone, fanning themselves with their wings to disperse the resultant odor, which can help the rest to find their way back from far away under adverse conditions.

A later researcher, Julien Francon, tried to reestablish the bee language view. He felt the other bees were too successful at finding food sources to be searching based on nothing more than smell alone. His claims, however, were sufficiently exaggerated that very few people took them seriously. He held, for instance, that the forager bee could communicate not only the distance and direction of the source, but also other characteristics such as its color, the presence of dry or wet sugar, and the mode of entry into a box. Nonetheless, he admitted that "We have no idea by what mechanism, auditory, or vibratory, these communications are established."

In 1919 Karl von Frisch began studying European honeybees (especially black Austrian bees, *Apis mellifera carnica*). His initial interest was in directions other than communication, most notably the extent to which bees have color vision. In the course of exploring those matters, he evolved a technique for training bees to exploit a particular food source, based on leading them gradually to locations progressively farther from the hive. He noticed, of course, that the returning bees would often dance, and he detected an apparent correlation between the form of the dance and the food source.

The basis for von Frisch's initial hypothesis was the observation that the bees have two distinct kinds of dance. The first possibility is a "round dance": the bee dances around in a circle, alternately clockwise and counterclockwise. The second possibility is a "tail-wagging dance": the bee moves in a straight line for a ways, wagging her tail (about 13 times per second), then makes a half circle back to the beginning point, repeats the same straight line, makes another half circle in the other direction, and so on for several cycles. The specific properties of these dance types will occupy us later in this chapter. For now, we need pay attention only to von Frisch's original observation that the forager might perform either of two distinct dances upon her return.

It was immediately clear to von Frisch that the two dances were not chosen at random. When the bees coming back had visited his nearby sugar source, they did the round dance; when they came back (independently) from pollen foraging, they did the tail-wagging dance. He interpreted this difference as the key to the message conveyed by the dance: the choice of one or the other form indicated the type of food (nectar or pollen) that the forager had discovered.

Other researchers challenged this view, having observed round dances by returning pollen foragers, but when von Frisch repeated his experiments he got the same result and he took it to be correct. If that were the entire story, the communication system would not be all that fascinating. Subsequent research, however, converted this relative banality into something vastly more interesting, which has spawned several generations of detailed study.

During the 1930s von Frisch was one of Germany's leading biologists, which insulated him somewhat from criticism for his role in sponsoring Jewish graduate students. "Eventually, however, his high standing in academic circles could no longer shield him [from the Nazi authorities], and he was sent back to his family estate in Austria for the duration of the war." While there, he returned to the question of the bees' dance language.

A new view resulted when one of his coworkers, Ruth Beutler, had occasion to train some bees to visit a feeding station somewhat farther (about 500 meters) from their hive than any involved in von Frisch's earlier experiments. These bees did a waggle dance when they returned to the hive, even though they were collecting nectar. Obviously, food type was not the key to their behavior (or at least not the only one), and something else had to be going on.

The experiments yielded much more. When von Frisch set out food at several different distant stations, the bees recruited by a particular forager's dance did not simply find an arbitrary one of these. Most, in fact, came to precisely the station the earlier forager had visited. This led von Frisch to the conclusion that the dance provided relatively specific information about the distance, direction, quality, and nature of a discovered food source.

Von Frisch and his colleagues then set out to explore in detail the connection between the form of the dance and the location of the food source. An extensive, painstaking series of experiments eventually led to the description of the system in his book *The Dance Language and Orientation of Bees*. That account has attained the status of a scientific classic and is generally presented as established fact. Actually, though, somewhat more work was needed before the underlying assumptions could be taken as established. Now that those details have been largely filled in, we can see that von Frisch's theory was essentially correct—and one of the more remarkable discoveries that have been made about communication among animals.

What Are the Bees Doing?

Before proceeding further with interpretation of the bees' dances, let us lay out more of what is (and is not) understood about their behavior. The basic context that interests us is one in which various bees (including both foragers seeking new sources of pollen and nectar and those exploiting already discovered sources) come back to the hive at more or less random times. Other workers in the hive unload the cargo from those who have been successful and perform appropriate tasks with it (these matters do not concern us here). The ease with which such "unloaders" are found is a source of information to the forager about the overall demand for her product in the hive at the time.

The returning foragers may or may not do their dances, depending largely on how abundant a food source they have been exploiting. Since many bees return during a given period, several "competing" dances may be going on at the same time. Other bees (those who have not just returned from successful foraging) consider the various possibilities on offer, eventually focusing on whichever one appears the most profitable.

The question of what determines whether a potential recruit will be attracted to a particular dance is not trivial. The load that an individual bee can carry (in terms of its energy value) is small, and the journey to the food source involves a significant expenditure of energy. It is therefore necessary for the good of the community that individual bees optimize their choices, not necessarily just following whatever dancing forager they may come upon first. Thomas Seeley describes a range of subtle experiments that work out the details of this sort of competition. The issues include what factors influence returning foragers to dance (or not) and which dancers win recruits. As a result of this work, the dynamics of this complex system are now fairly well understood, in terms of the economics of energy as this affects the hive. What matters to us is the recognition that information about the site being advertised by an individual dancer plays a major role in the process.

The complexity involved is complicated by the question of how the potential recruits extract information from the dance. For us, the answer is obvious. Bee researchers usually set up an observation hive with a glass panel and simply watch the bees. But bees do not have nearly the visual acuity of bee scientists; even if they did, they normally dance in the dark. Before we can conclude that some property of the dance we observe is also accessible to the bees, we must establish that they have access to it. Substantial mysteries remain to be solved in this domain.

Axel Michelson considers a number of possible modalities for information transfer in the dance and concludes the most likely is that the bees pay attention to patterns of vibration produced by the dancer in the surrounding air. Highly specialized sensory organs may provide the other bees with access to these sounds. The attunement of the response of these organs to properties of an ecologically important stimulus is typical of animals. We will see something similar in the specialized auditory sensitivity of frogs, which also matches the specific sounds that are important to the species. The rate at which the bee's abdomen vibrates during the tail-wagging dance is approximately 260 times per second, almost exactly the resonant frequency of the outer portion of a "listening" bee's antennae and also the frequency of maximum sensitivity of a special vibration sensor (Johnston's organ) at the base of the antennae.

Evidence has begun to accumulate that considerable information is transmitted in the waggle dance by vibrations in the honeycomb substrate, sensed by organs in the bee's legs. By integrating the data available from all six of her legs, a worker can derive specific information about the dancer's movements. This comb vibration apparently is critical: when it is absent, the number of watchers declines considerably. Experiments with mechanical simulations of dancing bees are also much less successful if the model bee does not contact the comb so as to induce such vibrations.

Some other bee species depend, to various degrees, on subsonic vibration (again, perceived by highly specialized body parts) and tactile contact. Although we have learned a great deal about what a bee knows about an-

Figure **4.1** The round dance: "The dancer is followed by those three bees
who trip along after her and receive the information."

other bee's dance (and how she comes to know it), there is much we do not
know in detail. We cannot really assume that our description of the dance is
closely similar to the way it might be manifested in the awareness of the bee.

Let us, however, take von Frisch's word for what is going on, describing the dances in the terms he (and nearly all subsequent researchers) employed. We can then match this description with the actual, observed properties of the food source, to determine what kind of information the bees
may be conveying and how.

In essentially all honeybees, some readily apparent gross differences in
the form of the dance provide an initial classification. The bees employ one
possible type, the round dance (Figure 4.1), when the food source is relatively nearby. What counts as "nearby" varies from about 3 to 80 meters,
depending to some extent on the subspecies. It is worth noting that this
value is "hard-coded" in the bee. If bees of one type are introduced to an
environment where the relevant distance is not the one to which they are
accustomed, neither they nor their audience will ever learn to "get it right."
The same is true for the more complex parameters of other dance types.

The form of the round dance is fairly simple. One bee dances in a circle,
reversing her direction after each circle is completed, for a time that depends on how rich the food source may seem. The information we can extract from this is that "a food source is present within some limited distance
from the hive." Apart from this basic message, potential recruits can also
determine what sort of food might be involved, because the returning forager is covered in the scent of her find, which is obvious to the organs of
smell of the others. The dance also indicates a rough estimate of the quantity of food available, manifested in the rate, agitation, and duration of the

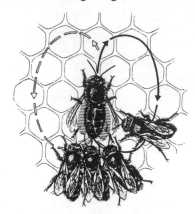

Figure **4.2** The tail-wagging dance:
"Four followers are receiving the message."

dance. (As I have noted, these dances are also used to indicate potential
nesting sites during swarming. When I refer to a food source below, this
other possibility is equally implied, in the appropriate context.)

The round dance does not, however, indicate anything more precise
about the distance and direction of the dancer's find than that it is somewhere within a limited circle. This is no great disadvantage, because the
area the bees must search is rather small. The other bees respond by going
out and flying around more or less randomly, looking for the kind of flower
that smells like their source.

When food is farther away, the area to be searched would quickly become too large for efficient probing at random. It is in exactly this case that
the bee does the more complicated dance on which her fame chiefly rests:
the waggle or tail-wagging dance (Figure 4.2).

This dance has several component parts (Figure 4.3). First, the bee
runs in a straight line, with a particular orientation, waggling her tail at a
specific rate. Then she circles around to return to the starting point and repeats the tail-wagging run, after which she again circles back to the start,
generally in the direction opposite to that of the previous return run.

The properties of the waggle dance are correlated with a number of
characteristics of the location of the food source. First, there is the orientation of the tail-wagging run portion. In the most basic of conditions, the
dance itself is done on a vertical surface within the (dark) hive, so we can
characterize the orientation of this component with respect to an imaginary line pointing straight up. When the tail-wagging run is also directed

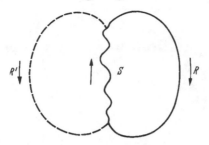

Figure **4.3** Pattern of the tail-wagging dance. *S*, waggling segment;
R and *R'*, return runs (as a rule alternately to right and left).

straight up, it corresponds to a food source that is in the same direction
from the hive as the sun. When the tail-wagging run goes straight down,
that corresponds to a food source in the opposite direction from the sun.
Orientations in between correspond to other possible directions, always on
the basis of the relation between the direction of flight and that of the sun.
Figure 4.4 illustrates three such possibilities.

In some instances, the bees may dance in a hive that is partially exposed, and in which the sun (or a light source interpreted by the bees as the
sun) is visible. In that case, the dance is no longer oriented with respect to
the vertical. Instead, the orientation of the dance is based on a line pointing
directly toward the sun itself (or its stand-in). This fact has been the basis
of some particularly ingenious experiments, as I discuss below. The bees
can still convey the direction of a food source even when forced to dance
on a horizontal surface, with no information about the position of the sun.
Under these circumstances they orient themselves with respect to the force
lines of the earth's magnetic field, treating this axis as the analogue of a path
toward the sun.

Apart from the direction of the food source, the dance can also supply
information about how far away it is. The duration of the waggle run conveys the distance: the farther away the food source, the more time is devoted to this portion of the dance. Different bee subspecies have different
"conversion factors" relating time to distance, but a commonly cited figure
is about 75 milliseconds of wagging for each 100 meters of distance between
the hive and the food source. Michelson shows that matters are not quite
this simple: the wagging run includes both tail wagging and sound emis-
sion, and both are apparently evaluated for the overall interpretation. In
experiments where the sound and the wagging do not correspond to the

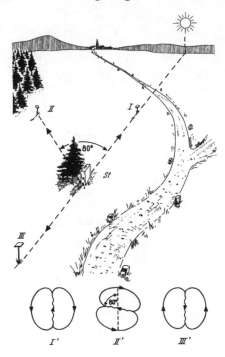

Figure 4.4 Three examples of the indication of direction on a vertical comb surface: *St*, beehive; I, II, III, feeding stations in three different directions; I′, II′, III′, the corresponding tail-wagging dances on the vertical comb.

same distance, the bees are recruited to a distance representing the longer of the two.

As with the round dance, the general level of excitation of the dancer and the overall duration of dancing behavior correlate with a rough estimate of the quality of the food source (the amount of pollen or nectar to be found). When the source is particularly desirable, the dancer shortens the time between the end of the straight waggle run and the beginning of the next run, and she hurries to repeat the part of the dance that conveys specific information.

Discussion of honeybee dancing is usually confined to the round dance and the tail-wagging dance, two forms whose information content is fairly clear. Several other patterns of dance-like movements can be observed in the hive, however. One of these is the trembling dance, which von Frisch described as follows:

At times one sees an odd performance by foragers who have returned home. "While they are running about irregularly and for
the most part at a slow pace over the combs, their body, in consequence of twitching movements of the legs, continually makes
quivering excursions forward and backward, to right and left. As
this activity proceeds, they run about on four legs with the forepair,
also quivering and twitching, held aloft in the position a begging
dog holds his forepaws." The "trembling dance" may subside after
only a few minutes, but may also continue longer, even for several
hours, I think it tells the other bees nothing. They do not bother
about bees doing the trembling dance and when by chance they
make contact with such individuals they are not impelled thereby
to any definite performance.

Thomas Seeley has shown that von Frisch was wrong in denying any
communicative value to the trembling dance. Close observation shows that
returning foragers perform this dance when they fail to find enough other
bees to unload the nectar or pollen they have brought back. Furthermore,
tremble dances by the frustrated bees have the effect of recruiting more
workers to the unloading task in times when particularly large quantities
of food have become available.

All of these matters can be (and indeed, have been) quantified, and a
human observer armed with the appropriate data can proceed directly to a
food source recently exploited by a dancing worker bee. Furthermore, the
other bees seem to pay close attention to the dancer, following her around
as she dances, and may well fly out to exploit the same source themselves
shortly thereafter. Of course they may fail to follow up the dance they observe, based presumably on the range of alternative possibilities on offer in
the hive at the same time, and the energy-economic consequences of flying
that far for the available level of reward.

Surely it is obvious that communication is taking place here. The first
bee must be telling the others where the food is; they understand what she
is "saying" and make use of that information to go to find it. But have we
actually shown that? A number of basic questions remain:

- Does the original forager "intend" to communicate and thus to recruit
  others?
- Why does the dance correlate with the location of the food in the way it

does? That is, where does the behavior come from? Do the bees "learn" how to dance, or is the behavior innate?

- Is the information communicated in the dance really what leads the others to the food? Or is the connection between the properties of the dance and the location of the food source something human scientists can determine, but not the other bees? Is their subsequent flight to the same source guided in some other way?
- If it is indeed the dance that provides the information, which of its properties are meaningful to the other bees, and how do they extract the information they need?

Despite the seductive nature of the correlation between dance properties and location, we cannot simply assume that the dancing bee "intends" (at any level) to communicate anything. The dance might be nothing more than an automatic response, an expression of the bee's internal state after returning from her foraging.

To see this, consider another potentially informative insect behavior pattern. A human observer can get quite an accurate idea of the temperature outside by listening to a cricket's chirping. Count the chirps during a period of 14 seconds (some say 15), add 40, and the result is the temperature in Fahrenheit. Despite the straightforward information conveyed, there is no reason to interpret the cricket's behavior as "intending" to communicate the temperature to anyone. The rate of chirping depends on the insect's internal state alone. It is logically possible that the more eloquent dance of the bee might have the same character.

It is consistent with this "deflationary" story that the dance system is completely innate and involves no learning at all. This is apparently the case with most bee behavior. For instance, bees that smell of oleic acid (a product of decay released by dead insects) elicit a preprogrammed behavior pattern. Other bees presume that any bee emitting this aroma is dead. They pick up the presumed corpse and carry it toward the hive entrance, eventually ejecting it. This phenomenon is fairly easy to understand, since a large number of bees die within the hive at any time and must be removed. Yet even a bee that is alive and kicking will be treated in this way if dabbed with a bit of the chemical that produces the behavior in the other bees.

As for the potentially communicative parameters of the dance, several factors show the innateness of the system. One is the fact that bees raised in

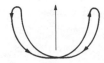

Figure 4.5 The sickle dance (*Sicheltanz*)

isolation, lacking any experience of dancing by other bees, will nonetheless dance correctly as soon as they are introduced to the hive.

Furthermore, the dance has "dialects." Some of these differences have to do with the details: different types of bees have different conversion factors of distance per waggle, for instance. Other differences are more structural: Italian bees (as opposed to the Austrian bees studied most closely by von Frisch) have a third pattern, the "sickle dance" that they perform for food sources at an intermediate distance.

The sickle-dance pattern illustrated in Figure 4.5 conveys only the direction, not the distance, of the food source. It is actually a sort of degenerate form of the tail-wagging dance, with the waggle run reduced to nothing but the orientation of the overall pattern.

How can we tell that these dialects are not a matter of learning (like the differences among dialects of a human language), but rather a matter of biological necessity? When bees with a different "waggle factor" are introduced into a hive (after appropriate precautions to keep them from being killed), their dances are interpreted in the wrong way and recruits arrive too far away or too near. Under these circumstances neither the dancers nor their audiences ever learn either to modify the dance or to interpret it correctly. For any given bee, the relation between dance properties and location of food is cast in stone as a result of the dance's genetic basis.

Hybrids of Austrian and Italian bees do the sickle dance if and only if they physically resemble the Italian bees. The dance form must therefore be inherited along with other genetically determined properties. Crucially, what is inherited is a specific system, not just a capacity to acquire *some* system within a given range. In contrast, when an Austrian and an Italian human have children, those children can learn either German or Italian (or Potowatami, for that matter), depending on which language they are exposed to early in life. In particular, we cannot predict which language they will learn on the basis of whether they look more like their father or their mother!

Two questions above (Is the dance really what conveys the location of the food source to the others? If so, what aspects of it, and how?) concern the extent to which we should take von Frisch's original account completely at face value. Before addressing that issue, I consider the extent to which a system with the properties attributed to the bees' dance language qualifies as a language, and its relation to systems such as that of human natural language.

Is This System a "Language"?

Recall von Frisch's first hypothesis that the bees communicate a choice between nectar sources and pollen sources by the difference between the round dance and the waggle dance. If that were the full story, we would be no more tempted to call this a language than we would, say, a car's turn signals. Surely it is of the essence for a language to provide a framework within which we can articulate (and comprehend) an unbounded range of possible messages. The power of human language derives from our ability to use it to say (and understand) things that are novel. If the set of messages were limited and fixed in advance, this possibility would not exist. A bee language in which only two things could be said would be qualitatively (as well as quantitatively) different from human language.

When we see how the system really works, this objection seems to dissolve, at least in part. The round dance, it is true, sends only a single message ("Nectar/pollen/nest site is nearby"). Similarly, the tremble dance can only convey the message "Food is here! Come and get it!" The waggle dance, however, conveys its message of location in terms of three parameters (distance, direction, and quality), and each of these parameters can in principle have *any* value within the range of possibility. "Food at 1,200 meters, 15 degrees off the sun, quality 4/10" is a message distinct from "Food at 700 meters, 135 degrees off the sun, quality 6/10," and so on. Since each of the three parameters (or at least distance and direction) can in principle vary continuously, literally an infinite number of distinct messages can be conveyed.

In mathematical terms, a parameter that ranges over a continuous interval can take on as many values as there are real numbers, a value represented symbolically as $\aleph$. A human language such as English, in contrast to the bees' system, produces new messages not by choosing points on a continuous interval but rather by combining discrete elements in new ways. As a result, the number of possible sentences in such a system is the same as the number of *rational* numbers, or $\aleph_0$. Because $\aleph_0$ is less than $\aleph$, the communication system of the waggle dance is—in principle—actually richer than English, in terms of the sheer number of potentially distinct messages. Bees are not capable of arbitrarily fine discrimination in perception and production of the various elements of the dance though, so this discussion of continuous-valued parameters is undoubtedly irrelevant in practice, at least as the basis for estimating the expressive capacity of the dance system.

In at least one sense, then, this system is quite rich. It can convey an infinite number of possible messages. On the other hand, one might object, it is also quite impoverished. Bees can talk about only one thing, namely food. This statement is a minor simplification, since bees that are swarming, and thus thinking of moving, use the same system to indicate a nice hollow tree or other site in which to set up a new hive. But one could still reduce the conceptual content of all messages to "location of what we are all looking for right now."

It is probably unfair to treat this feature as a limitation of the communication system per se. Arguably, it is really all that bees have on their minds, and thus whatever expressive failure it represents is inherent in bees' imagination, not uniquely in their communication system.

Many communication systems found in the animal world are far more limited than that of the bees, in that they can express only some fairly small finite number of distinct messages. What is formally the same message may have different force in different contexts, but bird calls and vervet monkey or prairie dog alarm calls convey *in themselves* one of a small number of messages. The bees' system, in contrast, correlates a range of different locations that is unbounded (in principle) with formally distinct dances, one to one.

The systems of bee dances and of human languages both achieve an infinite range of possible expressions. As we will see in Chapter 8, human languages manage this by combining elements from a vocabulary of fixed signs syntactically in varying ways. *Bee stings boy* and *boy stings bee* are quite different messages in English that result from distinct combinations of the same basic elements. Not only the words themselves but also the manner of their combination contribute to the meaning of a sentence. We might ask, then, whether the bee dance system (specifically, the tail-wagging dance) displays a syntactic organization as well.

The dance does indeed involve combining discrete elements in a fixed sequence (first a particular tail-wagging run, then a clockwise semicircle, then another tail-wagging run, then a counterclockwise semicircle—repeated as warranted). This sequence is part of what gives the dance its meaning; the same elements combined in some other way would not yield a meaningful tail-wagging dance. This system is unlike the syntax of, say, English, because combining the elements of the bee's dance in different ways could not serve to convey different messages.

The bees' system has a property that human languages do *not* have: continuous variation along its component dimensions. Human languages have discrete, not continuous, vocabularies (whatever a "continuous" vocabulary might be). Words are simply distinct from one another, rather than one shading gradually into another. Both bees and humans have an unlimited number of different things they can say, but that unboundedness arises from different aspects of the two systems. Our unbounded knowledge has a finite base: our vocabulary, or the words we know (surely finite, if large: perhaps fifty thousand words), taken together with some finite system of grammar.

The way speakers of a natural language derive an unlimited number of messages from the finite means at their disposal is through the use of what Steven Pinker calls a *discrete combinatorial system:* "A finite number of discrete elements (in this case, words) are sampled, combined, and permuted to create larger structures (in this case, sentences) with properties that are quite distinct from those of their elements." As Chomsky, Pinker, and others have pointed out, discrete combinatorial systems are rather unusual in the natural world. The genetic code and the immune system also involve small inventories of elements whose different combinations yield a vast range of possible expressions, but "most of the complicated systems we see in the world . . . are *blending systems,* like geology, paint mixing, cooking, sound, light, and weather."

A blending system involves a small number of fixed dimensions, where a particular message corresponds to a choice of points on corresponding scales. For the tail-wagging dance, the relevant dimensions are the orientation of the waggle-run portion and the rate of waggling. Another blending system is the representation of color on a computer monitor as some combination of values of three basic colors (red, green, and blue, for example). The number of possibly distinct messages in such a system is limited only by the system's capacity to distinguish finer and finer values for the continuous parameters that underlie it. In principle, the number of colors has no limit. For any two colors that differ only in the amount of red they contain, we can always use an intermediate amount of red to make a new color that is not the same as either of the others—although whether or not an observer will be able to detect the distinction depends on the acuity of the perceptual system.

Although both bee dances and human sentences form systems that can express (in principle) an unlimited number of distinct messages, they go about this in completely different ways. The "syntax" of a tail-wagging dance contributes nothing to its meaning; it only defines what constitutes a well-formed dance. In contrast, the syntax of a sentence is crucial to the way its component words contribute to a greater whole. The bee chooses each of the components of the dance along a dimension whose values shade continuously into one another, whereas each of the words of a sentence has a separate identity and sense.

Having said this, I must immediately point out that the way we use language in concrete acts of speaking involves more than the contributions of vocabulary and grammar. One important additional system is known as *paralanguage:* the gestural and emotional accompaniments to speaking, such as loudness and tone of voice, pitch (to some extent), rate of speech. These aspects of speaking we can vary continuously, to indicate, for example, an unlimited number of degrees of excitement by correspondingly varied loudness or pitch.

In many languages—perhaps a majority of those spoken in the world—distinctions of pitch are employed to distinguish lexical words in the same way that differences such as that between *p* and *b* serve in English. In Mandarin Chinese, as often noted, a syllable such as *ba* can have four different meanings, depending on whether it is pronounced with (a) a high, level tone [*bā*, "eight"]; (b) a mid-rising tone [*bá*, "pull out"]; (c) a falling-rising tone [*bǎ*, "grasp"]; or (d) a high, falling tone [*bà*, "dam"]. The tonal distinctions here are categorical. Pronouncing *ba* with a pitch contour midway between *bá* and *bǎ* does not denote an action midway between pulling out and grasping! The pitch distinctions in such a system are a part of language, not paralanguage.

Suppose the rest of language worked the way paralanguage does. Then we might have a system in which, for instance, we could pronounce a sentence like *Fred ran toward the lake* with varying pitch on the word *ran* to indicate Fred's speed, or varying loudness of *toward* to indicate how far he ran. Language does *not* work that way, though. What we can express by varying pitch, loudness, tone of voice, and other paralinguistic factors is our level of excitedness, anger, seductiveness, and the like rather than actual content.

We get new messages by combining different words and/or the same words in different combinations.

By and large, paralinguistic aspects of speaking express our internal state, not propositional content. We have considerable control over the extent to which our internal state is accurately reflected in this way, but basically paralanguage is much less voluntary than language. It is less a part of the communicative intention of the sender, although still fairly communicative from the point of view of the receiver. What we communicate through our facial expressions or body language is typically even less voluntary, but still rather expressive.

Distinguishing language from paralanguage is not easy. Structurally, the difference is dramatic. Language is a discrete combinatory system, whereas paralanguage is a blending system. Since the bees' system has the formal properties of a blending system, we might entertain the possibility that the dancing is like paralanguage in being a largely involuntary expression of the animal's internal state, rather than an intended message with external reference. I examine this question briefly at the end of the chapter, where I consider problems and refinements in the interpretation of the dance.

One other major difference between human language and the dances of the bees concerns the *arbitrariness* of the signals themselves (see Chapter 2). All of the aspects of the dance that could convey meaning do so *iconically*, in that there is a close correspondence between the structure of the signal and what it indicates. The level of agitation of the dance conveys the forager's level of enthusiasm for the food source; the dancer's orientation with respect to a fixed axis (defined by gravity, the sun, or, in a pinch, by the earth's magnetic field) conveys the orientation of flight; and the duration of the waggle run conveys the duration of the necessary flight. In contrast, English words such as *good, bad, north, south, near, far* have no nonarbitrary relation to the notions they convey.

Problems and Refinements

The rather precise nature of the correlation between properties of the dance and location of a food source is beyond any doubt. At the very least, the forager's dancing plays a role in recruiting additional workers that will exploit the same source. Nonetheless, some questions remain. Von Frisch (and we) can extract information from the dance that is directly related to the location of the food source. Can (and do) the other bees do something comparable? If so, how much information do they obtain? From what source? And how? In fact, how do we know the other bees are actually finding their way to the food source as a result of *any* specific information gleaned from the dancer, as opposed to using their sense of smell or sight, or simply following her, or even random luck? The more we learn about the intricate structure of the dance, the more we are tempted to believe in its directly communicative nature, and some of the alternatives seem implausible. Still, these aesthetic considerations do not in themselves constitute evidence.

Do Bees Get Their Information from the Dance?

Evidence suggests that some of the information human researchers can extract from the dance is not available (or at least acted on) by the bees. For instance, research has established that while doing the round dance, the bee produces sound during just that part of the circular pattern where she is oriented toward the food. Other foragers ought, then, to be able to fly directly toward the source. Bees recruited by these dances search the nearby area at random, however, showing that they do not make use of this clue. How much more of what science finds in the dances is similarly undetected or irrelevant from the bees' point of view?

The issue is not entirely abstract. In fact, in the history of research on these matters, at least one important challenge has been mounted to the dance-language theory. Given von Frisch's nearly legendary status, together with the obvious appeal of the account he offered, alternatives have not been much publicized in the wider community. Yet a different picture is defended by Adrian Wenner and Patrick Wells. Although current evidence argues fairly clearly that their view is incorrect vis-à-vis that of von Frisch, the factual and methodological issues they raise are relevant, both to an understanding of bees and to the broader study of animal cognition and communication.

Wenner and Wells pointed out that virtually all of von Frisch's work was devoted to showing a systematic correlation between the properties of the dance and the location of the food. But that left open the extent to which the other bees make use of the information. And it is surely important, in characterizing any proposed communication system, to show not only that it provides information that *can* be extracted as a message, but also that the recipient of the communication actually *does* extract that information.

Wenner and Wells suggested that the actual way recruits find their way to the food is through odor. The cues include specific pheromones released by foragers who happen upon a rich food source, faint trails of smell the initial forager may leave behind in the course of her return to the hive, and the general odor of the relevant neighborhood, detectable from the forager herself during the dance along with the smell of the food source. Bees have an extremely keen sense of smell, so the possibility that odor is what leads them to return to the source exploited by the dancer cannot be excluded.

One piece of evidence invoked to show that the dance cannot be as informative as von Frisch thought is the fact that the recruits take several minutes to arrive, much longer than their speed of flight requires. Wenner and Wells's experiments suggested the predominant role of odor and the ineffectiveness of the dance.

Other experiments have shown a more nuanced picture. In some circumstances, particularly strong odor cues suffice to overrule anything else. In fact, Wenner and Wells used extremely strong-smelling solutions in their experiments, so this effect probably was the determining factor in the followers' behavior.

In other (more normal) circumstances, however, it must be the dance, not odor, that leads recruits to the food source. For instance, where different dances indicating different food sources are taking place at the same time, the vast majority of the recruits arriving quickly at a given food source are those that watched the corresponding dance. This suggests, at a minimum, a connection between the specific dance and the recruits' behavior, but it does not exclude the possibility that olfactory cues of some sort play a role.

A series of more interesting experiments showed that it really must be the properties of the dance that guide the recruits. Bees fed a mild dose of a neurotoxic chemical dance more slowly than normal, for example. Recruits that have watched the dances of these bees arrive at an incorrect location. They appear to interpret the dance in terms of the distance it *seems* to indicate, rather than the location the dancer had actually visited. We do not find an alternative here to the assumption that the bees are going to a location communicated (perhaps incorrectly) by the dance.

One particularly ingenious experiment is considered to have settled the issue. A bee that can see the sun—or a light source interpreted as the sun—will dance with an orientation based on a line to that light source. The effectiveness of this cue is controlled by the overall level of light, as perceived by three simple eyes (called *ocelli*) at the top of the bee's head, distinct from the insect's main compound eyes. If the ocelli of a forager are painted over, the bee will not perceive certain levels of an artificial light source as "really" being the sun, although normal bees will indeed interpret it in this way.

Making use of this phenomenon, we can set up a situation in which the forager bee dances in a pattern whose axis is determined by gravity (because she does not see the "sun" as such), while her audience interprets her dance on the basis of an axis established by the artificial light source. In this way the forager is induced to "lie" in the sense that the message conveyed to the others is not a correct description of the location of the food source. The recruits go where the dance points, not where the dancer "intended" — showing clearly that they are getting their directions from (their interpretation of) the dance.

Other researchers have found different ways to establish the informational role of the dance. One of these is the development of an artificial, mechanical bee, whose dance movements can be programmed by a human controller and used to explore the relation between dance patterns and the behavior of recruits. Thus far it has proved rather difficult to fool the bees completely with such a contrivance, but these experiments have achieved at least limited success. Adding sound improves the extent to which the robot bee attracts recruits, and still greater improvement can be expected when a way is found to make the model induce appropriate vibrations in the honeycomb substrate. While the mechanical bees are less effective than real bees, their gyrations can direct a certain number of recruits to arbitrarily chosen locations. Since the mechanical bees have never been there, the recruits obviously cannot be finding the spot by smell or other cues left behind.

We conclude, therefore, that the dance does seem to be the main source of information for recruits about the location of a food source, at least in the absence of special conditions such as an overwhelming odor cue. Precisely what aspect of the dance structure is informative is under active investigation, now that effective mechanical bees allow for controlled experimentation.

These results largely vindicate von Frisch's original picture. It is important to recognize, however, that such vindication was truly necessary. Von Frisch himself did not do (or even envision) the experiments that establish the facts, and until they were performed, something like the "odor trail" hypothesis remained a possibility.

What Do Bees Mean by "Distance"?

Does the dance also have the interesting property we referred to in Chapter 2 as *displacement?* It is sometimes claimed that displacement is unique to human language, but it is possible to argue that the bee's dance refers to something in the external world. The dance appears to represent the distance between the hive and the food source—a factor that is not itself present in the actual communicative situation.

The question here is whether the relation between the time spent tailwagging during the dance (together with sound output, another potentially relevant cue) on the one hand, and external distance on the other, is arbitrary and symbolic. It certainly looks that way, but things may in fact not be as abstract as is commonly assumed.

We can approach this question by asking what aspect of their experience the bees are reporting in the dimension of the dance we interpret as distance to the food source. One possibility is that this parameter is based on the duration of the flight made by the forager in going from the hive to the food. That is unlikely, however, because bees flying very different routes to the same food source will report roughly the same distance. Something more complicated than just the time spent in flight must be involved.

Von Frisch believed that the relevant variable determining the bees' report of distance was not time, but energy expended in flight. This interpretation seems to be supported by a number of experiments. For instance, bees flying with little weights attached, or with strips of tinfoil to increase wind resistance, indicate sharply increased "distances." Bees flying into a head wind, or up a steep slope, or in a wind tunnel similarly indicate longer flight paths than under more normal conditions. And in another series of experiments, bees were forced not to fly but to walk to food sources. Upon returning, the affected bees registered "distances" about twenty-five times the actual distance to the food. These results suggest that the bees' dances signify how much effort they had to expend to get to the food source, rather than the literal distance between it and the hive. This hypothesis in turn reduces the extent to which the bees can be seen to be referring to something external (a geographic measure) as opposed to something within themselves (their perceived sense of effort).

Figure 4.6 Honeybee in a tunnel with complex visual patterns

Newer work has shown that the energy expenditure theory needs re-vision. When bees are trained to a food source on an island, so that a sub-stantial part of their flight is over water, their dance differs depending on the appearance of the water as they fly over it. When the water is completely calm, they tend to report distances that are less than the "real" distance from the hive to the food. In contrast, when the water is rough and choppy, they register substantially greater distances, even though the actual dis-tance (and the amount of effort expended in flight) remains the same.

This result suggests that the bees' perception of "distance" might be af-fected by altering the animal's visual experience. A series of experiments reported in *Science* magazine involved forcing bees to fly through some or all of a tunnel with a complex visual pattern on its walls (see Figure 4.6). Even a relatively short flight in the presence of this elaborate visual texture led to dances reporting greatly increased distances. When the experiment-ers replaced the patterned tunnel with an identical one lined with uniform black and white lines oriented along the length of the tunnel, reported dis-tances went down sharply. It became clear that the experience of distance that contributes to the pattern of the dance is based (at least in part) on the extent to which the image in the bees' eyes changes during the flight.

What about the earlier results, which appeared to show a sensitivity to the amount of energy expended on the flight? In fact, these can be explained by the same hypothesis. When bees fly in heavy winds, or uphill, or when otherwise burdened, they fly much closer to the ground than under other circumstances. Similarly, when a bee is forced to walk in a low tunnel, she is also much nearer to a surface. The closer the bee is to the ground or other surface, the greater the density of changes in the visual pattern as she traverses a given distance.

What is evident is that the parameter of the dance that indicates distance to the food source does not represent a genuinely external property, but reflects the bee's subjective experience in flight. As a result, there is much less to suggest that the distance parameter in the bee's communication system involves genuine displacement.

The parameter of direction is somewhat more suggestive. Again, some experiments imply that the orientation of the dance may reflect the bee's internal state (as it results from her flight) rather than objective properties of the location of the food source in the world. Bees flying in a crosswind are apt to indicate the wrong direction, so perhaps the message of their orientation is not "This is the direction from the hive to the flower," but "This is the way the sun looked to me while I was flying." Although the difference may be subtle, the latter interpretation is consistent with a nondisplaced account of the message of the dance.

The "mistakes" made by the bees in these cases might really only show that they are not perfect. Other observations argue that it is not the effort expended in, or the subjective direction of, their flight that matters, but the location of the food. As an illustration, when bees have to fly some distance in one direction and then turn, or in general when they find food after random foraging flights, their dance indicates the direction from the hive to the food, not the direction of their flight.

The skeptic could still argue, in response to these facts, that the insect's subjective experience of her flight is not a simple and literal retelling of what happened, but rather is based on a kind of "dead reckoning" that integrates the various twists and turns of her path to arrive at a unitary internal impression. This integration might determine the shape of the dance, without the need to refer to any kind of cognitive map of the area.

Other considerations have been interpreted as showing that such cognitive maps do exist, however, and that bees have the ability to indicate points on such a map with respect to a fixed point (the hive), independent of the way they got to (or from) that point. These are subtle and fascinating matters, and I can hardly hope to settle them here. My goal, rather, is to point out just how complicated it can be to establish that the seemingly obvious properties of a communication system are what they appear to be.

In the end, what kind of communication system should we attribute to the bees in their dancing? Perhaps the dance is not a simple involuntary expression of their internal state, although it has the formal properties of human paralanguage (another gradient system that is expressive, rather than being primarily referential) and other evidence seems consistent with that interpretation. The extent to which it makes sense to call the dance intentional is unclear, but at a minimum the messages conveyed must indicate a rather elaborate and cognitively rich kind of internal state. The dance seems to be specific and communicative. Bee "language" is a system of iconic signals that convey meaning; formally, it is a system quite unlike human natural language. Nor is it acquired on the basis of an interaction of the organism's genetically determined nature with the environment, in the way human language is.

All of these matters will occupy us further in later chapters. They do not in any way detract from the fascinating nature of the bees' behavior, but they do make the contrast with human language quite obvious. And exploration of these topics helps us go beyond the simple notion of "communication," to clarify just what language is.

5

Sound in
Frog and Man

For years and years the Doctor had been patiently working on the study of insect languages . . . In fact, there was practically no branch or department of insect life that the Doctor had not at one time or another studied with a view to establishing language contact with it. He had built many delicate machines that he called listening apparatuses.

But in spite of a tremendous amount of patient labor, trial, and experiment, he had admitted to me that he felt he had accomplished nothing. So you can imagine my surprise when he came rushing into the dogs' dining room, grabbed me by the arm, and breathlessly asked me to come with him. Together we ran across to the insect houses. There, over the various listening apparatuses, he attempted to explain to me how he had at last achieved results—results that, he was confidently sure, would lead to his dream being realized.

It was all highly scientific and complicated, and I am afraid that I did not understand a great deal of it. It seemed mostly about "vibrations per second," "sound waves," and the like. As usual with him on such occasions, everything else was laid aside and forgotten in his enthusiasm.

—*Doctor Dolittle's Garden*

The preceding chapters have been concerned mostly with general questions about methods and underlying assumptions in studying animals. That discussion sets the stage for a more detailed exploration of the nature of human language and its relation to the communicative and other cognitive abilities of animals.

When we think of language, we think first of speech, of language as it is conveyed in sound. Chapter 9 makes the point that, even apart from ways of representing speech in another modality such as writing, language is not necessarily connected with speaking. But the link between the usual medium in which we experience it (sound) and the message itself (language) is a seductive one. Some have even taken sound to be part of the very definition of language, and it seems natural to begin our inquiry with a consideration of how sound functions in spoken language.

In a wider sense, we might ask, why sound? Especially since we know that linguistic form can be represented in another modality in signed languages, we might think it more or less an accident that humans have seized on noises they can make (and others can perceive) as a medium of communication. The question is more complicated than that, however. Human biology is rather precisely adapted to the way speech functions in language, even at the cost of some serious disadvantages in other regards.

To understand these matters, we need a bit of basic background about how different patterns of vibration of the air (sound) can convey different messages. We begin by looking at instances of auditory communication whose structure is simpler than that of human speech, namely frog vocalizations. On the basis of what we learn from these, we examine the way sound is produced and perceived in humans, and then move on briefly to the ways differences of sound are structured and utilized in the systems of natural languages.

Why Does a Frog Croak?

Frogs are among the best known and most studied of animal models for a number of biological systems. We have been willing to experiment on frogs in ways we would never consider for our fellow humans (or even other mammals), as generations of high school biology students can attest. In cognitive terms, frogs turn out to be quite exquisite specialists, because their neurophysiology involves systems that are specific to tasks that are pertinent to the lives they live.

One of the classics of my own scientific education was the work of Jerry Lettvin and colleagues on "What the Frog's Eye Tells the Frog's Brain." The visual system of the frog, it seems, contains fairly precise "bug detectors." These are neural circuits that respond to just those patterns of retinal stimulation that the typical motion of an insect within the frog's visual field might produce and that are connected to circuitry controlling the extension of its tongue. We may not know "what it is like to be frog," to paraphrase a famous essay by Thomas Nagel about the mental life of a bat, but whatever that may be, we know a vast amount about how frogs do what they do.

Communicative behavior is no exception, and one of the more intensively studied examples of vocal communication in the animal world is the calling of various species of frogs. We probably have a much more comprehensive picture of frog communication in all its aspects than of human linguistic behavior. An extensive literature has established in minute detail the nature of the sounds frogs make, the physiological mechanisms of sound production and perception and the relations between these and the neural and endocrine systems of the animals, and the significance of these phenomena for the frogs' behavior. Like the visual system, the auditory system of the frog is structured in a way that is closely connected to the specific use the animal makes of it.

Why *do* frogs make those calls that can be so much a part of the atmosphere of spring and summer evenings? The big reason, and the one on which science has focused its investigative lens, is mating. Central to this fact is that at the appropriate time of year, male frogs congregate somewhere and begin making calls to attract females.

As a dating strategy, this system works quite well: individual females of the species approach a calling male and mating takes place. Some males have better luck than others in attracting females, so there must be something nonrandom about the process. The females must be making active choices, and it is reasonable to ask what it is that lady frogs want in a mate and what information they use in making their choices.

In some species, choice of a prospective mate might well be made on the basis of estimating the relative abilities of various contenders to protect and provide for the female during the production and nurturing of progeny. Frogs do not make responsible fathers, though, and the female is going to be left completely alone with her fertilized eggs. She is under no illusions about this: her choice is determined knowing that all she will get from her mate is his genetic material. That is, there is no reason for a female frog to prefer one male over another except for the question of which one will produce the most viable offspring. From that point of view, there is a simple rule of thumb: bigger is better. To the extent she can, she chooses a large male over a small one.

Why, one might ask, do the male frogs gather together, instead of separating? Because the dramatic chorus resulting from the joint efforts of a large number of contending bachelors is more likely to attract females. The males do keep a certain amount of separation from one another, not only spatially but also acoustically. As one frog calls, another waits nearby for him to be quiet before producing his own call. To some extent, that is, they take turns.

If the only requirement for this system to work were a way for the males to make a noise that the females can hear, there would be no particular reason why the males would pay attention to one another's calls, or indeed hear them at all. But the male frogs perceive the calls their fellows are directing at the females so that they can space themselves and take turns. The balance is subtle: on the one hand, each frog takes advantage of the joint efforts of the entire chorus to make a maximal impression on the females, but on the other, each frog wants to attract females to himself and not just to the group. He thus has to act in a way that will make it possible for a female to detect and distinguish his stellar qualities and pick him out of the crowd.

Shifting our perspective, the importance to the females of hearing the calls is pretty obvious. They need to be able to identify males of their own species and select among them. In many species the acoustic properties of the call vary as a function of body size. At least one measure of viability is thus signaled in the sound itself: bigger frogs (who are likely to produce bigger, tougher offspring) have deeper voices.

Of course, even frogs don't think *only* about sex; their vocal and auditory systems have other functions as well. Perceptually, they must be able to hear what is going on around them—approaching predators and so on. Their hearing thus cannot be (and is not) limited to mating calls. But within the overall constraints imposed by the need to perceive the environment, the system is organized to respond particularly acutely to these ecologically essential vocalizations of their species.

In terms of production, too, frogs have calls distinct from those involved in attracting the opposite sex. Males commonly produce a quite different sound when they are angry—for instance, when another calling male intrudes too closely. In addition to mating calls, special signals may exist to *discourage* mating. South African clawed frogs (*Xenopus laevis*) have a sound they make when one male mistakenly tries to mate with another ("Get off me, you idiot"), as well as sounds made by the female when she is unreceptive to a male's advances ("Not tonight, dear, I have a headache").

The bulk of scientific attention, though, has been devoted to "advertisement calls," made in general only by the males. In many frog species, differences in the size and shape of the male and female larynx correspond to the sex-based differences in what the larynx is used for. Females do occasionally make isolated calls, and injecting them with testosterone can result in their calling more vigorously. The connection between the vocal–auditory system and the endocrine system has been thoroughly established: male sex hormones stimulate the brain regions that initiate calling.

On the receptive end, too, there is a connection between sound and the rest of the animal's biology. Obviously when the female hears the call, she is attracted. When a male hears another male call, he is affected as well. His genitals enlarge and he is likely to start calling. (Kelley and Tobias discuss these connections between the vocal–auditory system and the endocrine system in some detail.)

In broad terms, we have just characterized the vocal communication system of frogs. But we want to know more than that: just what sounds do they produce and how, and how are these perceived?

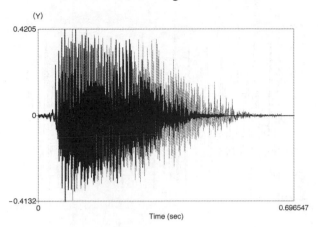

Figure 5.1 Waveform of an utterance of *boy*

The Structure of Sound

Before we can describe the frog's vocal and auditory apparatus (let alone our own), we need to say a bit about the physics of sound. An acoustic signal consists of a time-varying pattern of pressure in the air, and we can represent it graphically that way, as it might appear on the screen of an oscilloscope. Figure 5.1 shows an utterance of the word *boy* in this form.

It may be straightforward to say that this is what the acoustic event consists of, but the waveform itself is unenlightening: its structure is not at all apparent. Even when we narrow our attention to a small section in the middle of the *o* in this word (Figure 5.2), it is difficult to see any interesting structure in the sound.

Not only does the waveform fail to give us much information about the structure of the sound, it also encodes information that is not at all relevant to the way the sound is perceived. This will become more relevant when we talk about human speech perception, but for the present we can take it for granted that most of the information in the sound is carried by its *sound spectrum*. The spectrum of the vowel *o* in *boy* is given in Figure 5.3.

What Figure 5.3 represents is the fact that this sound contains varying amounts of energy (represented by values on the y-axis) at various frequencies (represented by values on the x-axis). This kind of decomposition of the sound is considerably more revealing than the waveform alone.

The notion of a sound's spectrum is based on the idea that certain sounds are maximally simple. These "pure" sounds correspond to the single tone produced in striking a(n ideal) tuning fork, which results in a pattern

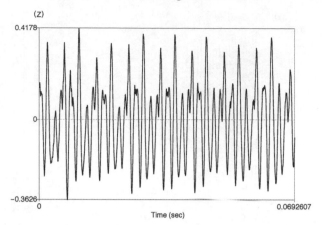

Figure **5.2** Central part of the *o* vowel in *boy*

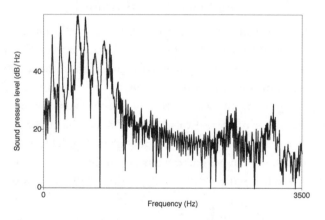

Figure **5.3** Spectrum of the *o* vowel in *boy*

of pressure variation in the surrounding air that describes a *sine wave*. If we combine several sine waves, the result is a sound whose pattern is more complex but which we can think about as made up of several individually simple components. And when we indicate the frequency and relative amplitude (strength) of each of these components, we obtain the spectrum of the complex sound.

What is the relevance of this discussion to the frog, an animal not known to employ tuning forks? Only this: It is a fundamental principle that the waveform of *any possible sound* can be regarded as constituted by adding together individually simple, pure sine waves, each at its own frequency and having its own amplitude relative to the others. The collection necessary to produce a given sound may be quite complicated, but there will always be one; and once we have described it, we have also described the sound to which the combination gives rise.

Actually, this is not quite true. In addition to giving the frequencies and relative amplitudes of the component sine waves, we must also describe their *phases* relative to one another in order to complete the description of a complex acoustic wave. Interestingly, the auditory perceptual systems of animals, including human hearing, seem to operate in such a way that complex sounds whose components differ only in phase generally sound alike. It is for this reason—and not just to avoid an additional layer of complication—that we describe sound spectra in terms of the two dimensions of frequency and amplitude alone. So Figure 5.3 characterizes the sound in Figure 5.2 in terms of a set of components at a number of frequencies and relative amplitudes (disregarding the phases of the components involved).

The task of analyzing sounds into a spectrum is intuitive, at least in part. It is easy to see that when we add several sine waves we get something complex. It is somewhat harder to convince ourselves of the opposite, namely that it is possible to decompose *any* complex form into the sum of a group of simple sine waves. That is true nonetheless, as long as the complex pattern repeats itself. And even a sound that does not seem to repeat itself can be analyzed this way, since it cannot go on forever. When it does change to something else, we can effectively pretend that it was going to repeat, and that what we heard was the first repetition.

The next step may seem as if we are pulling a rabbit out of a hat, but it is firmly based in the mathematics of functions like those that represent acoustic patterns. Let us assume that we do in fact have a pattern that repeats itself at some regular interval—say 100 times per second, or at a rate of 100 hertz (abbreviated Hz). For instance, the vowel *o* in the utterance of *boy* that is shown in Figure 5.2 involves a basic acoustic pattern repeated (more or less—the differences between one repetition and the next can safely be ignored) about 100 times per second. A result known as Fourier's theorem tells us that in the spectrum of this sound, *all* of the necessary sine waves will be at integer multiples of that rate (100, 200, 300 Hz)! These are *harmonics* of the *fundamental frequency* of the complex waveform. Although it is not obvious from the complicated curve in Figure 5.3, the spectrum presented for the single fundamental period shown in Figure 5.2 only has components at multiples of the 100 Hz fundamental frequency.

Perceptually, the fundamental frequency of a sound presents itself to us (and presumably also to the frog) as the overall *pitch* of the voice, and this is what varies in the frog's call as a fairly reliable indicator of his size. The precise distribution of relative strength of the harmonics of the fundamental frequency is what gives a sound its *timbre*.

On the basis of the facts just considered, we can—and generally do—treat complex sounds (that is, those that are not themselves pure tones) in terms of their spectrum, or the set of pure tones that we would have to add together to yield the sound in question. And when we say "This sound contains some energy at 1,000 Hz," what we mean is that one of the sine waves that would have to be included in its spectrum would have a rate of repetition of 1,000 cycles per second.

Armed with a way of talking precisely and explicitly about sounds, let us now plunge back into the pond with the frogs.

How Frogs Do It

Naturally enough (and fortunately for their ability to identify their conspecifics), different species of frogs have calls with somewhat different auditory structures. No single description will do for all. In part, these differences relate to species-specific variation in the structure of the larynx. (We will say little here about the precise mechanisms by which frogs produce their calls.)

The larnyx is a complex subject in its own right. For instance, since some species of frogs produce their calls under water, it cannot be the case that the function of the larynx is to control the flow of air by the same mechanisms the frog employs in breathing. It also appears that laryngeal activity results directly from specific neural impulses to specific intrinsic muscles of the larynx—as opposed to human speech, which involves laryngeal vibration due to a combination of aerodynamic and myoelastic factors.

What we need to know is that the frog's larynx, a valve-like structure between the lungs and the oral cavity, is primarily responsible for producing acoustic energy with particular temporal and spectral properties. It is that acoustic structure with which we will be primarily concerned.

The simplest kinds of calls are those of the *Hylidae* species. The call of, say, the spring peeper (*Pseudacris crucifer*) is a simple, repeated tone burst, in which all the acoustic energy is concentrated in one frequency region, generally around 2,800–3,000 Hz. Figure 5.4 shows the waveform of this call.

As we have already seen, the waveform does not reveal much about

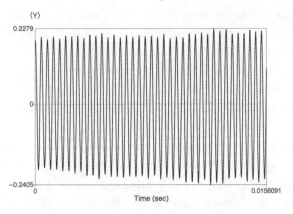

Figure 5.4 Waveform of the spring peeper's call

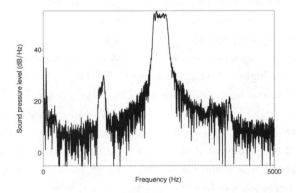

Figure 5.5 Spectrum of the spring peeper's call

the structure of the signal. To understand that, we need to look at the spectrum of the call. As Figure 5.5 illustrates, this call contains a concentration of energy at around 2,800 Hz, with a less prominent concentration in the neighborhood of 1,700 Hz. The production of such a call seems straightforward, but the response to it on the part of the perceptual system is somewhat more interesting. When we look at the frog's auditory system, we find that it is structurally rather complex. In particular, the inner ears of amphibians contain two distinct organs that produce nerve impulses in response to sound vibrations. One of these, the amphibian papilla, is sensitive to a range of relatively low frequencies — below about 1,200 Hz. The other one, the basilar papilla, is typically sensitive to higher frequencies, in the range 2,000–4,000 Hz.

Different cells in the amphibian papilla respond to different frequencies, and the frog can thus distinguish sounds in this range on the basis of the distribution of energy across their spectra. The cells in the basilar papilla all tend to be tuned to roughly the same frequency. Sounds within the range to which this organ is sensitive will result in neural activity that varies almost exclusively as a function of the amplitude of the energy within that range.

In the case of the spring peeper, the basilar papilla in the auditory system of the female is tuned to about 2,800–3,000 Hz. Not coincidentally, this is just the frequency that predominates in the male's advertisement call. The frequency that predominates in the calls of the biggest and most viable among the males, 2,900 Hz, is more or less optimal. The smaller the frog, the higher his voice, and thus the farther the concentration of energy in his call from the center of the range to which the female's basilar papilla is tuned. As a result, even if a smaller male puts as much energy into his call as his larger neighbor, he will not be heard as well.

The male auditory system involves a basilar papilla whose optimal frequency is a bit higher, perhaps around 3,400 Hz. In part, the difference between males and females in this respect follows from the nature of the basilar papilla, which is a fixed, passive resonator. Its size is a function of body size, and since in this species males tend to be smaller than females, the resonant frequency of their basilar papilla is higher. As a result, the male's basilar papilla response is close enough to the dominant sound in the calls of his fellows to hear them — but not so close that he goes deaf in proximity to the rest of the chorus.

The spring peeper's system of vocal communication is about as simple as we could imagine, and the connections between the way it works and the organism's biology are not in doubt. A somewhat more complicated system is that of the bullfrog (*Rana catesbeiana*). Robert Capranica explored the vocal system of this animal in a series of classic studies that were among the very first to establish a tight connection between the acoustics of the frog's vocalizations and the neurophysiology of its perceptual system.

Figure 5.6 illustrates the bullfrog's croak, which is obviously more complex acoustically than of the spring peeper's call in Figure 5.4. The spectral representation in Figure 5.7 shows this added complexity. The bullfrog's calls involve not just a single region of acoustic prominence, but a fundamental frequency and a number of its harmonics, with two general regions of prominence. One of these is relatively low (about 200 Hz) and the other

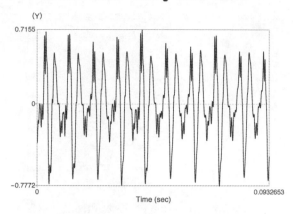

Figure 5.6 Waveform of the bullfrog's croak

relatively higher (about 1,500 Hz), with smaller amounts of energy distributed across the higher frequencies.

The frog's auditory system, whose overall organization is the same as
that of the spring peeper (and other amphibians), is closely attuned to the
structure of this call. These frogs are much bigger than spring peepers
and their basilar papillae are correspondingly tuned to lower frequencies:
around 1,500 Hz, in fact. Also, the responsive region of their amphibian
papilla is best at about 200 Hz. The two-part structure of the sound and the
two-part system of the amphibian ear are almost optimally matched, with
close correspondences in frequency between the production and perception systems.

Still more specialization of the perceptual system needs to be considered, however. The neurons that respond to signals in the amphibian papilla
of the bullfrog are subject to "two-tone suppression." What this means is
that if two close (but not identical) frequencies within the range of the amphibian papilla are both stimulated, the presence of the second tone *inhibits*
the response to the first tone. Bullfrog calls are definitively organized into
two distinct frequency regions, one within the range of each part of the
female's auditory apparatus. The system is built to prefer calls of exactly
that sort: if energy is present in an intermediate frequency region that corresponds to neither part of the standard call, the "two-tone suppression"
effect cancels out the low-frequency energy and reduces the perceptual salience of the overall signal. If something does not sound like a male bullfrog's croak, the female does not hear it as well.

Finally, the auditory neurons of the bullfrog can do "pulse train syn-

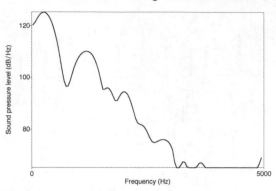

Figure 5.7 Spectrum of the bullfrog's croak

chronization" with low-frequency incoming signals. That is, if the frequency of a component in the signal is low enough, the relevant neurons do not simply emit a signal that indicates the presence of energy at that frequency; they emit pulses that are synchronized with the signal. The animal can then detect the fundamental frequency (of which this particular component may be a harmonic). Recall that the fundamental frequency is likely to be an indicator of the gross size of the animal, just as the pitch of a human voice tends to vary to some extent as a function of body size. Her auditory system therefore gives the female a way of assessing this characteristic in a potential mate fairly directly. The system is somewhat more sensitive than that in the spring peeper, where overall amplitude of basilar papilla stimulation is all that is available as a cue to body size.

The advertisement call is not something a frog picks up from experience, by paying attention to what other frogs are doing. There is no reason to believe that experience plays any role at all, in fact. Frogs raised in isolation call perfectly naturally without any opportunity to observe their fellows.

More important for our purposes, everything about the organization of the animals' ears and sound production systems is specialized so that male frogs make precisely the advertisement calls that female frogs like to hear. And the female frogs can determine from what they hear both the conspecific nature of the caller and aspects of his relative viability.

For very few species are the physiology and neurophysiology of the communicative system, as well as its function, so well understood as for frogs. That system of communication is not some adventitiously structured behavior on the animals' part that could readily have been otherwise (under

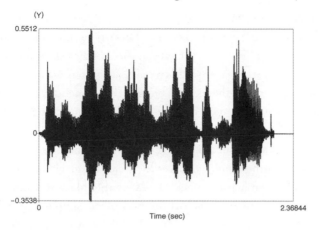

Figure **5.8** Waveform of a human utterance:
"I'm a really, really, really big frog!"

the influence of different training, perhaps). On the contrary, frog vocal communication has a structure that is deeply rooted in the frog's biology. This biological basis of communicative behavior will be a recurring theme in what follows, and of course is likely to be tied into the possibility that communicative mechanisms may be species specific and not merely the outcome of general cognitive faculties.

The Sounds of Speech

Any given species of frog has at most a small number of distinct calls (in the limiting case, just one), and I have said little about how these are actually produced. The central property of *human* language, however, is that we can produce an unbounded range of possible utterances (even though the ones we use for a specific purpose, such as that for which the frog's advertisement call is designed, may be limited to a few clichés). We therefore need to look into the way humans produce these various signals, after which we will consider the process by which they are apprehended and perceived.

How Humans Do It

A signal such as that in Figure 5.6, representing a bullfrog's advertisement call, has considerable internal structure, as we have seen. Still, a corresponding representation of an instance of human speech (as in Figure 5.8) is much more complicated.

Although the frog's call consists of a single basic element (the croak), repeated many times, human utterances can be regarded as made up of a variety of basic sound types or *phonemes*. This usage causes a certain amount of disgruntlement in linguists, for whom the word *phoneme* has a precise — and somewhat different — sense. However, as long as we are careful not to take seriously any of the theoretical baggage it has acquired over the years, it is as good a word as any to denote our presystematic sense of the basic sound types of a language.

According to a rich set of language-specific principles, phonemes combine with one another into words, which can themselves combine into a limitless set of possible utterances. A reasonable place to begin the exploration of this system is with the production of the simplest speech sounds, the vowels. Imagine a single vowel (say [a], the vowel of American English *top*), prolonged in steady-state fashion for a certain period of time. We can divide our analysis into two subparts: events at the larynx, and the structure of the supralaryngeal vocal tract.

The larynx serves essentially as a valve, regulating the flow of air from the lungs. It has two folds (sometimes misleadingly called the vocal "cords") that the speaker can bring together to close off airflow or can separate to allow it. In the production of a vowel like [a] the vocal folds open and close repeatedly and rapidly (somewhere between 80 and 500 times per second, depending on the pitch of the voice). The result is a series of short puffs of air, and that is what provides the acoustic energy in the sound.

Assuming (as we are) that everything holds steady, these puffs come at a constant rate: let us say 100 times per second, or 100 Hz. Recalling Fourier's theorem, we know that the acoustic spectrum of the sound produced must consist of this basic rate (the fundamental frequency) and any number of harmonics (multiples) of that frequency. We may find energy at 100 Hz, 200 Hz, 300 Hz, and so on — but nowhere else. In fact, the way the larynx works results in relatively even distribution of energy across a wide range of harmonics.

Laryngeal vibration thus provides us with a source of acoustic energy, which is the same for all *voiced* sounds (sounds in which the vocal folds vibrate). Therefore it does not tell us how [a] differs from [i] and [u] (roughly the vowels of American English *beet, boot*). These three sounds are distinguished, in fact, by the way the vocal organs above the larynx are positioned.

The larynx provides the basic source of energy (in vowels, at least), but to affect the surrounding air, that energy has to pass through the vocal tract, which acts as a filter. As a function of size and shape, the vocal tract reinforces some frequencies and attenuates others. What comes out can be obtained by "multiplying" the spectrum of the input energy source by the filtering function of the vocal tract. The regions where the vocal tract responds strongly are accordingly well represented in the output signal. These are called the *formants* of the sound.

Different vocal tract shapes will result in different patterns of resonance. We produce different sound outputs from the same basic source of acoustic energy by manipulating the position of the tongue, the lips, the jaw, and the velum so as to change the overall shape (and thus the resonances) of the vocal tract. Different speech sounds, naturally enough, are the result of variations in the way we position the organs of speech. That much is obvious, and the way it works is that differing resonances of the vocal tract, corresponding to different positions of the articulators, shape the output sound in different ways.

In Chapter 2 I mentioned the view that there are no actual "speech organs," that everything we use in speech is there for some other purpose. The whole system, from lungs to lips, is basically built to serve the needs of respiration and glutition, and gets pressed into service for speech production as an add-on function. Although this claim has some foundation, it is not as valid as it may seem.

The form of the vocal tract and the organs we employ in speech have evidently been determined over the course of human evolution in much larger part by our needs in speaking than might appear. When we compare the human vocal tract in Figure 5.9 with that of a primate relative such as the chimpanzee as in Figure 5.10, we notice several differences.

Overall, the base position of the human larynx is much lower in the throat, with a large cavity (the pharynx) above it, as shown in Figure 5.9. The placement of the velum, which separates the oral and nasal cavities, is also rather different, as is the placement of the tongue body with respect to the cavities of the vocal tract.

Together these differences yield considerable advantage to humans for speaking: the pharynx provides a substantial, flexible resonant cavity whose size and shape can be easily manipulated by changes in the shape of the tongue, making it possible to articulate a wide variety of sounds. The chimpanzee vocal tract, in contrast, is incapable of producing most of the vowel sounds of human languages ([a], [i] and [u] are beyond its capacity, for example), as well as many consonants (such as the velar sounds [k],

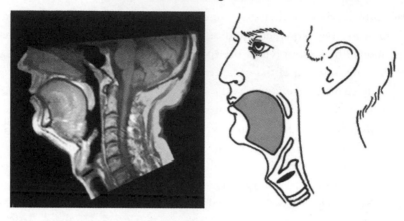

Figure **5.9** Midsagittal MRI of *Homo sapiens*.
The tongue body is shaded in the diagram at right.

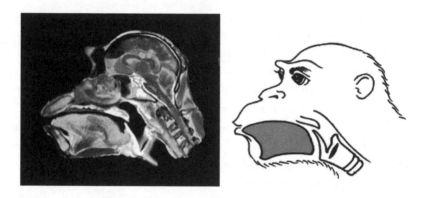

Figure **5.10** Midsagittal MRI of *Pan troglodytes*.
The tongue body is shaded in the diagram at right.

[g]). The position of the velum also makes it impossible for the chimpanzee to produce distinct nasal and nonnasal sounds ([m] as opposed to [b], for example).

The advantages of the human vocal tract, though, come at a nontrivial cost. In particular, a substantial part of the human vocal tract serves simultaneously for the passage of air to and from the lungs and for the passage of food and drink to the digestive tract. Chimpanzees can use their velum to separate the two channels, making it possible to breathe while eating or drinking; humans who try to do so will choke. From the point of

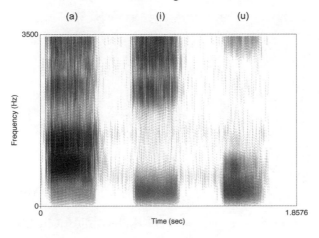

Figure 5.11 Acoustic spectra of [a], [i], [u]

view of respiration and glutition, the human system is poorly designed, requiring adroit and precise coordination of various valves to prevent severe failure.

What we see in the course of evolution, however, is that the human vocal tract has evolved from something rather like that of the chimpanzee (as found, for instance, in some reconstructions of the vocal tract of Neanderthal man) to its present shape. In Chapter 11 we shall learn that the differences between the modern human vocal tract and that of other animals, perhaps including early hominids, are not as absolute as is often assumed. Nonetheless, the basic shape shows distinct evidence of having been adapted for the special purpose of speech. If we think of the vocal organs as primarily structures for breathing, eating, and drinking, their shape is absurd. In contrast, if we see these developments as adapting existing structures so as to make them into organs of speech (while preserving, as well as possible, their original functions), everything makes sense.

The energy source at the base of the vocal tract, then, will give rise to signals with different spectra as a function of the tract's resonances. Figure 5.11 gives a graphic representation of the difference among three basic vowels ([a], [i] and [u]).

The precise differences among these vowels are not my concern here. What matters is that the three have different acoustic properties, as a function of the various vocal tract shapes that operate as filters on essentially the same energy source.

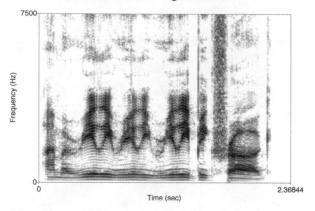

Figure **5.12** Spectrogram of "I'm a really, really, really big frog!"

The nature of the graphic representation in Figure 5.11 requires some explanation, however. The *spectrogram* is the most common way of presenting the acoustic structure of a time-varying event (such as a speech utterance). A simple spectral chart such as Figure 5.7 tells us the amplitude (one dimension) of the energy present at each possible frequency (the other dimension) at a single point in time. In contrast, a spectrogram is a *three*-dimensional picture that represents the relative intensity of energy at various frequencies as a function of time. A spectrogram of the utterance in Figure 5.8 is given in Figure 5.12.

For comparison, Figure 5.13 is a spectrogram of the series of four bullfrog croaks from which I extracted the spectrum in Figure 5.7, which represents one point in time during one croak.

The x-axis of a spectrogram represents the time dimension, the y-axis the frequency of the components of the signal, and the darkness of the impression (or sometimes the color, or other indicator such as a sort of topographic map) represents the amplitude of the energy present at a given frequency and point in time.

Classically, a machine called the spectrograph would pass the speech separately through a set of filters, each tuned to a different frequency, in order to make a spectrogram. This procedure caused a current to pass through special electrosensitive paper, resulting in marks that were darker at points where more energy was present at the frequency of the corresponding filter. Phoneticians used to get serious headaches from the ozone and other by-products given off during a prolonged session of spectrogram making, but nowadays the same function is performed quickly and pain-

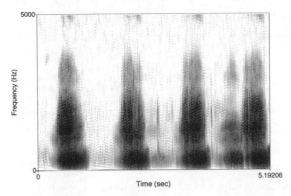

Figure 5.13 Spectrogram of a series of four bullfrog croaks

lessly by software available for personal computers. As we shall see, the spectrogram is not only a more useful way for scientists to represent speech visually, but also remarkably close in its structure to the way the peripheral auditory system deals with an acoustic signal in organizing its presentation to higher-level cognition.

When we move beyond simple signals like steady-state vowels to study consonants and sound combinations in running speech, matters quickly become much more complicated. The complications are primarily quantitative rather than qualitative, however, at least from our perspective here. It remains the case that the process of speech production can be seen as composed of (a) the generation of a source of acoustic energy, and (b) the filtering of that energy on its way to the air outside the vocal tract, as a function of variations in the acoustic response of the vocal tract itself.

Some variations in the sound that emerges from our mouths result from variations in the source. Changing the rate of vocal fold vibration, for example, changes the perceived pitch (or tone). Most of the acoustic variation that corresponds to distinctions among the phonemes of our language, however, results from changes in the configuration of our vocal organs. The marvelous gymnastics we perform with the tongue, the lips, the velum, and so on produce a correspondingly varying acoustic result. We can represent that output spectrographically, as in Figure 5.12.

Perhaps most interestingly, we can also recover the dance of the vocal organs when our perceptual system operates on the incoming acoustic signal. This is a crucial step in the process by which a listener extracts the information that a talker is offering, and we turn now to the process that makes it possible. Human auditory perception is rather more elaborate than that of female frogs tuning in to the chorus of their male conspecifics, but in both cases the systems are rather precisely adapted to the tasks they have to perform.

The Nature of Speech Perception

What goes on when we perceive speech? In particular, what is the process by which we interpret sound as language and particular acoustic events as specific linguistic utterances? An important part of the answer is provided by the insight of Alvin Liberman and his colleagues that "speech is special." Something happens when we perceive speech that is distinct from what happens when we deal with other auditory stimuli.

Interpreting acoustic inputs as speech seems a simple matter of translation. Apparently we want to do something like this: As the incoming waveform arrives, divide it into successive chunks corresponding to the successive phonetic segments. Then we ought to be able to analyze each chunk with respect to a set of "property detectors" and thus arrive at an identification of the phoneme represented by that particular chunk.

This program seems hopeful. For instance, all vowels that are articulated with the tongue body high in the mouth (like [i], [u] or German [ü] — as opposed to vowels such as [a], [e], [o] for which the tongue body is relatively lower) have a very low first formant, meaning that the lowest region of energy concentration in their spectrograms, like those of [i] and [u] in Figure 5.11, is lower than for others. We could, accordingly, build a detector for the property "high tongue body" that looks for a very low first formant. If we were to build such a detector for each of the *distinctive features* that linguists identify for the phonemes of a language, we should be able to apply the set of feature detectors to the successive waveform segments and emit their outputs as a phonetic transcription. Implemented in neurons, it ought to be a working system of speech recognition.

Designing the feature detectors might involve some engineering problems, but the process ought to be fairly straightforward. In fact, it is essentially the program that researchers began to pursue in earnest in the early 1950s.

The analysis of speech in this fashion, as a sequence of discrete sounds each distinguished by some particular set of acoustically identifiable properties, sounds like an obvious approach, given the way utterances are put together. If it is, though, we might wonder why it has taken so long to develop usable voice recognition software.

The answer certainly is not that no one has attempted to build an automatic speech recognition system along these limes. A voice-operated typewriter has been a sort of holy grail for a major segment of the electronics industry since at least the 1950s. We would think such a program would be easy, even trivial, despite problems in determining, for example, which spelling from among *pear, pair,* and *pare,* corresponds to the input. Something unanticipated, though, must make this problem a lot harder than it looks.

At the beginning of the twenty-first century, after many decades of effort, we are at last beginning to develop speech recognition software that more or less works, under favorable conditions. A general solution still eludes us, however. Virtually everyone involved would agree that the way speech recognition is achieved in current systems bears very little resemblance to the process that takes place in the human listener.

The problems in understanding and modeling human speech recognition arise (in large part) because speech is actually far more complicated than suggested by our view of it as a sequence of discrete sounds. A number of factors obscure the connection between a message construed as a nice, neat string of consecutive sounds and the acoustic reality that is its implementation by a human talker.

Recall that our suggested procedure began by chopping incoming speech signals into a series of discrete intervals. However intuitive that may sound, it is actually quite impossible: real running speech involves constant changes that provide no objective basis (either in acoustics or in the sequence of articulatory events that give rise to sound) for dividing the signal in specific places. Once we move beyond the artificial case of isolated, steady-state vowels (as in Figure 5.11) to that of complete natural utterances (Figure 5.12), we no longer have a way of picking out the discrete chunks of sound to which the proposed property detectors might apply.

This is not just because everything in speech articulation moves relatively smoothly and continuously, providing no telltale starts and stops that would allow us to delimit successive sounds from one another. Many sound types have acoustic correlates that are *in principle* inseparable in time from those of the sounds we think of as preceding or following them. Stop consonants like [b, d, g], for example, in words such as *Bob, bag, dog,* correspond not to separate intervals of sound preceding or following the vowel, but rather to precise patterns of change in the formants that define the vowels themselves, in their onsets and offsets.

Thus, when I say [bɪg] (*big*), *no* part of the resulting sound is the [b] or the [ɪ] or the [g] alone. At every point, information is present that is crucial to the identification both of the vowel and of one or both of the consonants.

As if this complication were not enough, it is also the case that speech is heavily *coarticulated*. That is, at any given moment in the production of an utterance (say, while I am focusing on making the vowel of the word *big* in Figure 5.12), the position of my vocal tract is not simply a function of a single sound. At this point the lips, the tongue body and blade, the state of the velum allowing (or preventing) the passage of air into the nasal cavity, and so on, are all functions not just of the vowel ([ɪ]) but also of the sounds that were produced just before and of those that are about to be produced.

At the beginning of the utterance, the fact that a nasal consonant [m] is coming up immediately after the initial diphthong (the [aḭ] of "I") means that the velum is already lowering during the production of the [aḭ], resulting in a slightly nasalized vowel. The labial articulation of the [m] requires that my lips are already being brought together during production of the [ḭ] part of this diphthong, resulting in a slightly rounded articulation (similar to that of [ü]). In each repetition of "really," the [r] preceding the first [i] vowel requires a specific position of the tongue that persists to some extent into the vowel, resulting in an articulation that is slightly different from that for the second [i] vowel, which is preceded by [l]. And so on, and so on, and so on . . .

Coarticulation has at least two distinguishable consequences, each of which makes it even harder to imagine dividing the speech signal into a sequence of discrete segments. On the one hand, it means that at any given moment the acoustic event is a product not of just one of the sounds that constitute an utterance, but also of several others—sounds that "precede" and others that "follow" any particular sound. And on the other hand, it means that the cues for identifying any one of these sounds are not localized in a single interval of time. Instead, they are spread out over a much longer part of the speech event, at least in principle—and, research confirms, in practice as well.

As Liberman stressed, language is *encoded* in speech. The structure of the signal reflects what we might represent in much more transparent form in a writing system as a sequence of letters; but the relation between the actual signal and such a representation is not at all simple or transparent. By and large, this complexity is completely hidden from the listener. We hear speech as a sequence of sounds, separable and coming one after another.

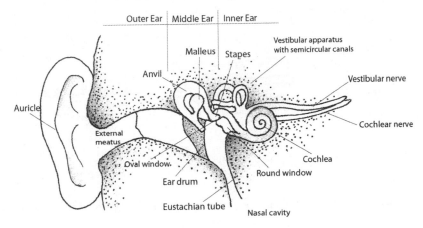

Figure 5.14 The human ear

Somehow we decode the complex acoustic event and extract its structure as linguistic units without being in the least aware of how different in form these are.

The Role of the Ear

How do humans accomplish this feat? Speech recognition is still a major problem, from both a conceptual and an engineering point of view, but before we can address the cognitive structures that make it possible, we should first look at the apparatus that serves as the first link in the chain: the ear.

Earlier in this chapter I argued that we can regard the complex wave-forms of speech in terms of time-varying patterns of energy distribution across a range of frequencies. That gives us a representation of the speech signal in the form of a spectrogram, a visualization that is much easier to work with than the full waveform. But this is not simply a matter of schol-arly and scientific convenience: the spectrographic representation is closely related to the way the ear analyzes an acoustic signal.

Figure 5.14 shows the overall structure of the human ear. Vibration that reaches the outer ear is transmitted to the *cochlea,* a long, fluid-filled tube. The cochlea is the core of the interface between sound and perception, the place where external pressure variations are converted to impulses on the auditory nerve. It is curled like a snail, a fact that is irrelevant to the way it works; for convenience I depict it in Figure 5.15 as if it were straight-ened out.

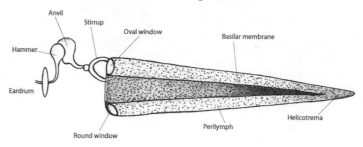

Figure 5.15 The cochlea, depicted as if unrolled

Down the middle of the cochlea runs the basilar membrane, which in turn is covered with the hair cells that make up the organ of Corti. Each of these cells is highly sensitive to pressure variations in the surrounding fluid, and their response is what translates the sound wave into stimulation of the auditory nerve. This neural activity indicates not only the presence of sound, but something of its acoustic nature: vibration at a given point on the basilar membrane tickles the corresponding hair cells, causing specific corresponding fibers in the auditory nerve to fire.

The key to the working of this system is the fact that different parts of the basilar membrane respond to energy at different frequencies, as a function of linear position along the membrane. When sound causes the fluid in the cochlea to vibrate, hair cells at different points along the length of the (coiled) cochlea are correspondingly stimulated, causing the corresponding neurons to fire (Figure 5.16). Just which hair cells are affected depends on the frequencies present in the sound waveform. Energy at high frequencies excites cells close to the base of the cochlea, and lower frequencies excite cells closer to its apex. The degree of excitation of the neurons connected to each of the hair cells, as a result, varies in a way quite similar to the variation in darkness of a single (horizontal) line on a spectrogram. The auditory nerve fibers are not connected one-to-one with the hair cells; rather, each one is stimulated by a group of locally contiguous cells. Each fiber's activity, then, represents the sum of activity in a range of frequencies. Activity in a particular auditory nerve fiber corresponds to the amount of energy present within that rather narrow range of frequencies over time.

All of this corresponds rather closely to the way the spectrograph works. The ear appears to extract information about the presence of acoustic energy as a function of frequency and amplitude (but ignoring phase), just as the spectrograph does. The ear, like the spectrograph, gets its infor-

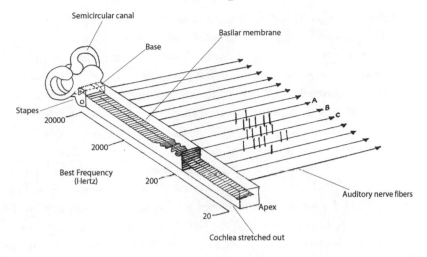

Figure **5.16** Conversion by the cochlea of sound to nerve impulses

mation in terms of a set of tuned filters, each of which averages the energy present over a relatively small subpart of the spectrum. It seems reasonable, then, to regard a spectrogram as being at least a very good first approximation to the representation of incoming sound that the ear provides (via the cochlea) for higher-level auditory processing.

A Motor Theory

We still need a way to use the information provided by the ear to achieve a cognitively real representation of speech (something like a phonetic transcription, for instance). As we have seen, the solution to this problem is anything but self-evident.

What looks like the most promising proposal is one that links our ability to sort out what we hear to our ability to speak. We are competent hearers, such an approach suggests, because we are also talkers. This is a *motor theory* of speech perception, in that it presumes that we hear speech not just as a series of abstract sound types, but rather as a sequence of speech gestures of a sort that we could, in principle, have produced ourselves.

The basic character of such a theory resides in the notion of what the categories are that function in speech. On this view, what happens when we perceive speech is not the assignment of labels to acoustic events, but rather an analysis of the input signal in terms of the articulatory gestures that could have produced it. The object of phonetic perception, that is, is not acoustics but a motor event.

As Liberman says, speech "informs listeners about the phonetic intentions of the talker," and a motor theory takes these intentions to be the production of certain gestures. What the hearer recovers from the signal is evidence about the gestural activity that gave rise to it, not the raw acoustic properties themselves. If this picture is appropriate, we would expect the categories of speech perception to be coherent in terms of the talker's gestures that produce a given set of acoustic results, but not necessarily in terms of the structure of the signal itself.

In support of this view, the sounds that sound the same may be quite different acoustically, and vice versa. We can observe that instances of the consonant [d], for example, are acoustically dissimilar, depending on the nature of the following vowel. What [d]s have in common regardless of their surroundings is that they are formed in the same way (making the tongue tip contact the alveolar ridge). Their similarity is based on their origins in the same gesture, regardless of the following vowel. The class of sounds that are thereby treated as "the same," however, would not be treated alike at all if we simply looked at acoustic events to see which ones were like which others.

The other side of this coin is the fact that some instances of [d] are actually much more similar, in purely acoustic terms, to some instances of [g] than they are to other [d]s. Not only do we class all of the [d]s together as instances of a single sound, we cannot hear (no matter how hard we try) the similarities between some [d]s and some [g]s. The two sound types are made differently ([d] with the tongue tip contacting the alveolar ridge, [g] with the tongue body contacting the back part of the palatal region), and they sound quite distinct, regardless of whether or not purely acoustic resemblances exist.

Another remarkable observation is that this effect of the origin of sounds in a talker's motor activity is apparently limited to sound *to the extent that we perceive it as speech.* According to the motor theory, when an auditory signal has the general properties of speech, we perceive it in a way quite distinct from the ordinary way of perceiving sound in general. This "speech mode" essentially recovers the articulatory activity (by the talker) that might have given rise to it, whereas general auditory processing provides information that is much closer to the acoustic properties of the incoming signal. Some special (and specialized) brain mechanisms must exist for perceiving speech, mechanisms that are different from those we use for nonspeech auditory processing.

Indeed, by now there is growing evidence for such specialized mechanisms, just as there is for specialized cognitive mechanisms in visual processing for specific tasks such as recognizing human faces. Paradoxically, though, some of the evidence that supports a distinctive cognitive mechanism for speech perception, one that operates by reference to the motor processes by which humans produce speech themselves, comes from studies of nonhuman animals.

In one case (to which I return in Chapter 6), the evidence for a motor theory is direct. We know that zebra finch song (like human language) is lateralized, in that different aspects of song structure are controlled by the two hemispheres of the brain. More interestingly, we have direct evidence for a motor theory of song perception in the zebra finch.

In this bird, we are able to record neural activity in brain regions that are known to be responsible for motor control, all the way down to the nerve controlling the syrinx (the principal song-producing organ in a bird), exactly when it hears conspecific song. Crucially, this activity is not present when the bird hears sounds that are acoustically different from finch songs. Studies of neuronal activity that we cannot possibly conduct (noninvasively) in humans are possible in birds, but perhaps if we *could* perform the corresponding experiments on ourselves we might uncover similar evidence for the implication of motor control mechanisms in human speech perception.

Another observation that makes a motor theory more plausible than it might initially appear is based on the discovery in the mid-1990s of "mirror neurons" in the brains of monkeys. These specific neurons can be shown to display similar activation patterns under two related circumstances. Mirror neurons are active when the monkey performs certain actions himself (grasping a food item with his hand, for example), and also when he observes the same action performed by another monkey. The actions of a human experimenter have the same effect, since our hands are apparently similar enough to those of a monkey to be perceived in the same way. Such a brain mechanism establishing an equivalence between motor activity and perception of the same activity in another is exactly what we need to construct a perceptual system that works along the lines suggested by the motor theory.

This view leads to the conclusion that phonetic perception is very different from auditory perception in general. Perhaps the two share virtually no structure apart from a dependence on (even competition for) the same class of input events. A basic difference between the perception of speech and of other sounds is in the nature of what is perceived. In audition in general, the perceived event, for which the acoustic signal provides evidence, is something "out there" in the world: barking dogs, breaking glass, a slamming door, or the notes of a clarinet—events not essentially related to the listener. Speech differs in that a listener is also a talker, and the gestures that produce the speech signal could also be made by the listener, with comparable acoustic results. The motor theory suggests that this relationship is not merely accidental, but is an essential fact built into a special mechanism we use to perceive speech.

A distinctive mechanism (or module) processes acoustic signals as speech very differently from the way auditory perception works in general. The result is a highly specialized and completely automatic analysis of a signal when the input is treated as speech. This analysis is not in terms of its acoustic structure, but rather in terms of the talker's gestures that could have given rise to it. As a result, different gestures that could yield the same acoustic result in some local context may nonetheless be perceived differently, to the extent that other information is available to support different interpretations. Similarly, different acoustic events that correspond to the same gesture in differing contexts will be perceived as the same despite the objective differences. The constants of perception, that is, correspond to those of articulation rather than to those of acoustics.

Further evidence for the idea that speech is perceived in terms of motor gestures derives from the so-called McGurk effect. In experiments in which the ear hears one thing and the eyes see an apparently synchronized talker saying something else, the result is a compromise. That is, the exact same acoustic event (say [ba]), produced by a talker who actually made a bilabial stop, may be perceived quite differently (as [ba], [va], [da]), depending on what visual image is present. It seems that listeners use all available evidence to interpret the talker's gestures and do not just assign a label to the acoustic signal itself.

The idea that speech perception is a distinct, specialized module, separate from general auditory perception, is supported by a number of facts. Some experiments utilize the paradigm of "dichotic listening" tasks, where different signals are presented (through headphones) to each ear of a listener. By varying what is presented to the two ears, it is possible to experiment with the auditory system's ability to synthesize two distinct channels of information into a unitary percept. An early result supporting the notion that speech perception is not the same as nonspeech auditory perception was the general finding of a right-ear advantage for speech as opposed to a left-ear advantage for nonspeech stimuli. Differential involvement of the brain's two hemispheres in the two sorts of task appears likely.

The following is an interesting dichotic listening experiment. We can synthesize signals corresponding to only some formants of a syllable consisting of a consonant followed by a vowel: say [ba] versus [da] or [ga] or [bi] or [di] or [gi]. If we restrict the information to what is present in the first two formants, the result will still be clearly perceived as [ba], [di], and so forth. When we separate out just a single formant, however, the result is no longer anything like speech: what we hear instead is a sort of "chirp." We can distinguish one chirp from another, but they sound completely unlike speech sounds. More important, the ones that sound similar are those that have something acoustic in common: "rising" chirps as opposed to "falling" chirps, for instance. When we play two of these chirps at the same time, presenting one to each ear, the chirp perception disappears, and we hear a single, unitary syllable (speech again). And now the signals that sound similar are not those with, for instance, rising chirps, but rather the ones that are perceived to have the same initial consonant, regardless of whether the perception of that consonant is cued by a rising chirp or a falling one.

A further refinement of this technique presents what seems to be solid evidence for the existence of a distinct speech mode in auditory perception. Only an experimental psychologist could love the manipulations involved.

Apparently when a very short portion of one formant from a three-formant representation of a syllable is separated from the rest, a unique "duplex perception" effect can be obtained. The starting point is a signal that could be either [da] or [ga], depending on whether the third formant is rising or falling at the beginning of the vowel. This piece of the third formant alone is very brief (about 30 milliseconds), but crucial to identifying the syllable. Culling it from the rest of the signal yields two separate sounds: (a) a signal that is perceptually ambiguous between [da] and [ga]; and (b) a short rising or falling chirp. Playing these to the two ears separately yields an astonishing result: the subject perceives *both* an unambiguous syllable of speech, perceived either as [da] or as [ga] depending on the nature of the chirp, and *also* a simultaneous chirp, sounding completely independent of the spoken syllable and not at all speechlike. This unusual manipulation of the stimulus manages to recruit both the speech mode and the nonspeech mode of auditory perception to interpret the same acoustic content. The chirp both determines the interpretation of the speech and is heard separately, as nonspeech.

Apparently the perceptual system must perform some sort of initial "triage" on incoming acoustic information, classifying some of it as speech and some as nonspeech, and referring each kind of data exclusively to the appropriate perceptual system. The duplex perception experiments involve tricking this initial sorting so as to activate both perceptual modes and allow some data to be processed by both independently.

All of which confirms, then, that the perception of speech in human listeners involves a distinctive cognitive process: the speech mode. Working independently of the cognitive processing that operates otherwise in the perception of auditory stimuli, this system interprets incoming acoustic information in terms of the motor activity on the part of a talker that might have given rise to the sound being heard. As such, it provides the listener with a way of getting at the linguistic intentions of the talker, and thus of circumventing the complex encoding involved in the implementation of speech.

We saw at the beginning of this chapter that the frog's systems for producing and perceiving sounds are quite precisely matched. The study of human speech reveals essentially the same match between perception and production, this time involving a specialized perceptual mechanism that operates separately from but in addition to the more general process of auditory perception. The neurophysiological bases of this unusual adaptation for speech communication are not as evident as in the case of the frog, but then again, the systems are vastly different in terms of the complexity of the messages being transmitted. In both cases the communication system and the manner of its implementation are tightly linked in a way that seems particular to the species.

We can of course argue about whether the perceptual system that links articulation and perception in the way we hear speech must be present in the organism at birth. It seems plausible that the capacity for forming such connections exists innately, though it requires experience to develop into a fully robust cognitive system. I argue in Chapter 6, in fact, that the babbling period in human infants provides a way of "tuning up" the associations between articulation and acoustics. Even if the overall structure of these associations is present at birth, it can use some honing by experience.

Speech perception as a special mode, based on assigning a motor interpretation to incoming information, raises many broader questions. For instance, other systems of skilled performance might similarly result in the emergence of a special motoric perception of appropriate stimuli. Music comes to mind, in the case of experienced instrumentalists. Regardless of whether it is unique to speech or an instance of a more general possibility, though, a distinctive, specialized, motor-based mode of perception provides the key to the way humans extract linguistic content from the complex (but ultimately highly efficient) code that is speech.

Phonology

To this point, we have described the ways in which two organisms (frogs and humans) produce and perceive sound for the purpose of communication. In the frog's case, once we have dealt with the mechanism of croaking and the organs (the basilar and amphibian papillae) by which the croaks are perceived, we are pretty much finished. In the human case, though, we are only beginning to approach the interesting part.

Hockett's notion of duality of patterning is a central property of human language. A small inventory of basic units (sounds) combine according to regular principles to form a larger inventory of meaningful signs (words, to a first approximation); and these combine according to a completely different system (syntax) to form an unbounded range of potentially distinct messages. There seems to be no system in nature other than human language that works this way—certainly frog croaking does not. Our account thus far covers the way the most basic combinatory units of a language are implemented and perceived in sound, but it says nothing about the ways in which these sounds are combined with one another to form meaningful units. Our story of sound in language would be seriously deficient if we concluded it here.

In linguistics, the study of sounds, their properties and formation, and the perceptual cues that allow listeners to recover them, is generally known as *phonetics*, whereas the study of their organization within the systems of particular languages in the formation of higher-level linguistic units is called *phonology*. Practitioners of each have long tended to be somewhat dismissive of those in the other field. The early phonologist Prince Nikolas Trubetzkoy famously suggested that phonetics is to phonology as numismatics is to economics, while otherwise sensible phoneticians regard phonological arguments as mere artful conjuring with no empirical basis. Both are wrong about the triviality of the other's pursuits, but it is probably in the nature of scientists that neither are likely to change their minds soon.

Among other things, phonology includes characterization of the particular selections that different languages make from among the sounds that fall within the human language capacity. For instance, English has at least a few extremely rare sounds, seen from the perspective of the world's languages. The *th* sounds in *thin, this* are found in only a small subset of the world's languages; and the vowel of words such as *bird* for many speakers of American English is incredibly rare. This is true for those who do not drop the *r* after a vowel, as do people in many parts of the northeastern United States. Most speakers of American English who do not drop the *r* in this position do not pronounce a real consonant there either, but instead modify the preceding vowel. This gives the vowel in *bird* a distinctive quality, written [ɚ] by phoneticians.

Conversely, there are many sounds that other languages use and English does not, such as the palatal *gl* of Italian *sbaglio* (mistake), or glottalized consonants such as the [k'] in Georgian *k'atsi* (man, person).

Some sounds are commonly employed in the languages of the world, others much less so. We can generalize about the nature of these sound inventories, and one of the tasks of the phonologist is to find these regularities, for they are part of the nature of language. Within the limits of what is phonologically possible, each language has its own specific system, which is part of what makes it the language it is.

This much is obvious to anyone who learns a foreign language, but the same sounds can also be put together according to quite different principles in different languages. To see this, we need first to notice that the set of phoneme combinations that are *possible* words in a given language is not the same as either (a) the set of *actual* words in the language, or (b) the entire set of possible combinations of phonemes of the language. The point is fairly intuitive: if we read in the instruction manual for a gadget we have just bought that the *flitch* should be left in its factory setting, our main concern will be where to find this object, whereas if we read the same thing about the *chlfi*, we will wonder about the obvious typographical error. *Flitch* (phonetically, [flɪč]) is a combination of English sounds that does not happen to correspond to any existing word, but it could; *chlfi* ([člfɪ]) contains exactly the same phonemes, arranged in a way that could not possibly be a word of English. This difference is nothing ordained by a hypothetical Academy of English Usage, but simply follows from what every speaker of the language knows about it.

Even when the phonemes involved are essentially the same in two languages, a combination that constitutes a possible word in one may be quite impossible in the other. For instance *mgla* is a Russian word (haze; gloom, darkness), but it could not be a word of English. *Fact* is a real English word, but not possible in Italian, where words cannot end with a cluster of consonants such as *-ct*. On the other hand, Italian *sbaglio* begins with a cluster ([zb]) that cannot occur at the beginning of an English word.

For world records in the category of consonant clusters, Georgian is perhaps the leader: *gvprts'kvnis* ("he is bleeding us, financially"), obviously not possible in English, is a real (monosyllabic!) word. Even here limits exist, limits specific to Georgian. For example, *sp* and other words that begin with the sequence *sp* are not possible Georgian words. Part of knowing a (particular) language is knowing these limitations on how sounds can combine to make the signs we use as words. Because the limitations can vary considerably from one language to another, they cannot simply be a matter of what the human speech production system is capable of articulating.

Phonology is also responsible for describing the ways in which the basic units of sound are related to phonetic reality, which ones count as different and what replacements occur under specific circumstances. Thus, in English the consonants *p, t,* and *k* are sometimes followed by a short puff of breath before the next vowel (aspirated) and sometimes not. The *p* of *pot* ([p^{h}at]) is aspirated, while that of *spot* ([spat]) is not. As English speakers, we notice no difference between these sounds until it is called specifically to our attention, because aspirated and unaspirated *p* do not contrast. In other languages aspiration counts as differentiating words. In Punjabi, [pət] (honor) and [p^{h}ət] (split) are two different words.

In English, vowels are relatively long or short depending on the following sound. The [a] of *cob* is much longer than that of *cop*, for example, but once again we do not notice this distinction until it is pointed out, because the difference in the vowels is completely predictable from the following sound. In Tahitian, the same difference in vowels can distinguish words on its own. In this language a *pati* (with a vowel like that of English *pot*) is a mountain range or a lineup of men, whereas a *paati* (with a long vowel similar to that in English *pod*) is a kind of fish.

Every language has its own principles that determine which phonetic differences among words are potentially part of the very identity of those words, and which are predictable—and in instances where they are predictable, how to predict them. Thus, aspiration in English is predictable from a principle something like "If a *p*, *t*, or *k* is at the very beginning of a stressed syllable, not preceded by *s*, it is aspirated—otherwise not." Punjabi has no such principle, since aspirated *ph* and unaspirated *p* are as different as *p* and *b* in this language.

We talk about sound structure in these cases in terms of a small set of basic units (sounds, or phonemes) that can be combined in various sequences like beads on a string. We can describe each sound in terms of some collection of properties, like whether the lips are closed, whether the vocal folds are vibrating, and so on. This analysis of sounds in terms of their distinctive features is, as Ray Jackendoff stresses, a major discovery in the study of language, comparable in many ways to the discovery that substances in the natural world are made up of atoms.

Another part of a language's phonology concerns the accommodations and adjustments that occur in the form of words when meaningful elements are combined in a way that would lead to violations of the range of permitted sound combinations. For instance, the regular ending added to nouns in English to make their plural is a [z], though it is written *s*. In words such as *cats*, the [z] is replaced by an [s], for the combination [tz] is impossible at the end of an English word (notice that the name *Katz* is pronounced just like *cats* in American English). And in *bushes* an extra little vowel is inserted to give [bʊšəz], because neither [šz] nor [šs] is possible at the end of a word in English.

These facts about the phonology of English do not just follow from what it is possible for the human vocal organs to do. In Surmiran Rumantsch (a language spoken in Switzerland), for instance, the ending of most plural nouns is *s*, similar to the English form. In this language the word *codesch* ([kodɛš], book) has as its plural *codeschs*, pronounced [kodɛšs]. No modification is necessary, since (quite incredibly, to a speaker of English) final [šs] is possible in this language.

There is much more to say about phonology; I have had time and space only to scratch the surface. The important point is that these facts and principles, aspects of the phonology of a language, are part of what speakers know about it. If we have not learned the phonology, we do not yet know the language, even if we are able to reproduce accurately a number of utterances in it. This point may seem trivial, but it will take on some importance in later chapters.

The Biology of Communication

We have seen that communication using sound can recruit systems at several levels. The very organs of sound production and reception constitute one such level, the one most amenable to direct study as a physical system. At another level, a communication system may involve specialized processing mechanisms in the peripheral auditory system, mechanisms that are sensitive to auditory stimuli of a specific character and that process these stimuli in a particular way adapted to the communicative function they may perform.

Sound is hardly the only mode of communication in which we find specialized processing mechanisms. Studies of olfactory communication in mice, for example, have shown that these animals have two distinct systems for the perception of odors. One involves a membrane in the nose called the nasal epithelium, which contains olfactory receptor neurons sensitive to a wide range of chemical substances in the environment. These receptors project to the main olfactory bulb, and from there to the animal's olfactory cortex. A second system involves a distinct structure called the vomeronasal organ. Located not far from the nasal epithelium in the mouse's nose, it is sensitive only to the pheromones that play a large role in the animal's reproductive and social behavior. Unlike the main olfactory system, the receptors in the vomeronasal organ project to entirely different regions of the brain, where information about pheromonal stimuli is processed separately from that about smells in general.

The biology of the mouse, therefore, involves a highly specialized system dedicated precisely to the processing of one subclass of olfactory stimuli. Pheromonal signals are specific in their properties, and they are extremely important to the animal in ecological terms. Thus, it makes sense that a specialized system should develop to process exactly these signals from the much broader spectrum of olfactory events.

Against this background, we can view the existence of a speech mode in human auditory processing as not fundamentally different from a bug detector in the visual system of the frog or the vomeronasal organ of the mouse. In the most peripheral case, that of the physical organs (such as the frog's croak production apparatus and the two sound-sensitive structures in this animal's inner ear, the vocal tract and the ear of humans, and the two olfactory systems in the mouse), there is no serious question that the structures we find are those determined by the organism's biology. They are aspects of its genetically determined structural identity as a member of its species. Nor is there reason to doubt that higher-level cognitive structures with similar functional specificity can also be rooted in species-specific, genetically determined properties of a species.

In the instances of the frog and the mouse, these matters can plausibly be reduced to the physical structure of the organism, at the neurophysiological level. Inasmuch as we have no trouble attributing the existence of bug detectors in frogs to frog biology, there can be no serious basis for denying the same status to the speech mode in human auditory perception. This is true even though the speech and nonspeech modes of auditory processing are distinct only functionally, not neuroanatomically (so far as we know). The similarly specialized mode of perception (for conspecific song) in the zebra finch *is* neurophysiologically distinct, at least in part, from other components of the bird's auditory system. Humans have a great deal of cognitive capacity within which distinct functions can be associated with the same tissue. Functional and anatomical specializations are more closely associated in the other animals we have considered, but that is no reason to doubt that all of these cases are grounded in species-specific biology.

When we come to phonology, a specialized organization imposed on sound material at higher levels of cognitive structure, things may seem different. After all, humans do not come with the phonology of their language in place. If they did, they would be unable to acquire the particular phonology of the language spoken around them. Interaction with experience is required, in the form of input data from a specific language.

In actuality, the situations are not that different. For one thing, it is likely that some interaction with experience is necessary to tune up the specific mechanisms of the speech mode. The babbling period in human infants may serve that purpose. And while phonology is particular to human natural language, we can find resemblances to systems in other organisms.

The structure of song in many bird species provides an interesting basis for comparison. At least some birds acquire a system of regularity about the sound structure of song, their species' mode of communication, through an interaction between innate mechanisms and concrete experience. Few would deny that the capacity to acquire such a system arises from the specific biology of the animals that possess it. There is no corresponding reason to doubt that the capacity to acquire the phonology of a language is just as deeply rooted in our own biology.

6

Birds and Babies
Learning to Speak

"Then I decided I'd try to learn the language of the birds. Clearly they had a language. No one could listen to their warblings and not see that. For years I worked at it—often terribly discouraged at my poor progress. Finally—don't ask me when—I got to the point where I could whistle short conversations with them. Then came the insects—the birds helped me in that too. Then the plant languages. The bees started me. They knew all the dialects. And . . . well . . ."

—The Moon Man, from *Doctor Dolittle in the Moon*

The Moon Man learned the language of the birds by dint of intensive listening and enormous amounts of repetition—the method sometimes applied in total immersion classes for human learners of foreign languages. When we study a second language as adults, we may approach it in this way or we may have a great deal of explicit instruction in the principles of grammar. Whatever we are doing, clearly it is not the same as what small children do during the period when their first language(s) can be said to be developing. We have no reason to believe that they need to work at it as hard—and as consciously—as we do. And there is little reason to believe that the end result of our language classes is the same as what develops spontaneously in the child exposed to the language spoken around her.

Just how *does* knowledge of a language develop in the child? In many ways, this has become the central problem of modern linguistics. Most theoretical proposals, for example, are accompanied by a story (sometimes a Just-So story, but a story nonetheless) about how languages having the properties attributed to them by this theory could possibly be acquired from the sort of data available to the child. What does the child bring to the learning task, such that the knowledge we observe in the adult can arise on the basis of plausible early experience? If we linguists could give a comprehensive answer to that question, we would have defined the very nature of human language.

In the present chapter we begin to address this basic question. We do so, however, on the basis of studying communication in another kind of animal, birds. The calls and songs of birds constitute an extremely well studied system, one for which the physiological and neurological mechanisms underlying production and perception are fairly well understood.

More directly to the point, birdsong shares an important property with human language. In three of the twenty-seven orders of birds, song develops in a way that requires interaction with early experience (rather than being entirely innate), just as human language development does. The lessons that can be learned from the way birdsong develops provide valuable precedents for study of the same questions in human infants. Not that the answers are exactly the *same*—language and birdsong are radically different systems in most respects—but strong enough similarities exist to make the comparison somewhat revealing.

Bird Calls

The ducks started honking to one another as they saw the dawn. It almost seemed as if they were exchanging signals as conversation of some kind because I suddenly saw that they somewhat changed direction following a leader, a single duck, who flew at the head of the V-shaped flock . . .

At length some signal seemed to be sent back from the leader up ahead. Because all the flock stopped and started circling and eddying away in the wildest manner. We had arrived over a wide, wide bay on the shoreline. The coast seemed low, and behind it were many ponds and lagoons. I could tell from the dizzy singing in my ears that my duck was descending—like the rest—in widening circles to the flat marshlands they had come so far to seek.

—Doctor Dolittle's Garden

Birds exhibit a great deal of interesting behavior and, as with most animals, much of what they do conveys information. Consider the case of the red-winged blackbird (*Agelaius phoeniceus*). This bird has red and yellow patches on its shoulders, which are obvious in flight but somewhat concealed when the wings are folded. Male aggressive behavior with respect to other males involves uncovering these "epaulettes."

If a bird's red wing patch has been painted over with black by an experimenter and the bird attempts to defend his territory in the usual way, the other birds will muscle into his terrain, even if he does everything right. The male bird's act of communicating his aggressive intentions through his display is crucial, and other birds assess the threat at least in part by its robustness. A painted bird has less to show off (though he has no idea of the change that has been wrought in his perceived masculinity). Obviously the display is communicative; we could even regard the bird's appearance, independent of his doing anything at all, as communicative. Still, we would surely not be tempted to think of this as "language" in any but the most general sense.

Much more interesting in the present context is the ways birds communicate using sound. We refer grossly to the two kinds of vocalization by which birds communicate with their fellow birds as *calls* and *songs*. The line between the two is not easy to draw sharply, in terms of either their form or their function, but it will be useful to have two categories.

For instance, I was out in the woods near my house in Connecticut recently, when I heard a flock of Canada geese (*Branta canadensis*) overhead. They were making a great deal of noise, all of it in the form of fairly brief squawks and whistles, short single notes with no real organization. These were calls, and virtually all birds produce them.

A given species has a characteristic inventory of calls, comprising a set of short simple sounds associated with particular events and activities. Many species, for instance, have an "aerial predator" call, which provides conspecifics (and sometimes other species as well, though this is presumably quite inadvertent) with a warning that an eagle, hawk, or similar predator has been sighted. Others, such as the Canada geese I heard, produce flight calls (which we could gloss anthropomorphically as "I'm taking off," "This way, guys," "I'm going in for a landing," and so on).

The main characteristic of a species' set of calls is that they appear to constitute a limited, closed inventory of discrete messages. The system cannot be extended at will to convey novel messages. Accordingly, there is no creativity or open-endedness in the messages themselves. In particular, calls cannot be combined by some sort of discrete combinatory system (see Chapter 4) to form more complex messages.

The calls all seem to be expressions of internal states in the bird that produces them, rather than symbolic expressions (like human words) that refer to things in the world. This does not mean that there is no relation between a call and things in the world: for instance, a rooster's likelihood of giving a food call depends at least in part on the presence of a hen in the environment, among other "audience effects." Still, we would not want to say that the food call *refers* to the audience—only that the chicken's internal state reflects his awareness of other chickens in his environment. An alarm call of some kind can be produced with greater or lesser urgency, depending on the animal's degree of agitation, and we will see that other circumstances can modulate the frequency or intensity of calling. It is the closest bird calls get to the character of blending systems, and it is not very close.

In contrast, songs essentially serve one intertwined set of functions. Fundamentally, songs are an expression of territoriality. Since it is mainly males that defend territories in the avian world, males tend to be the ones that sing—though in some species, such as the red-winged blackbird, females may sing too to compete for their own territory. The male's song sends a dual message. To other males, it serves as a territorial claim and a warning ("This Land Is My Land!"). To females it is supposed to be attractive ("Hey, Baby!"). The basis of the attraction is in part the singer's assertion of his territory, though the performance may convey other information about the degree to which he is a robust and generally fit mate.

Bird calls usually differ from songs in their form, typically being much less elaborate. In general, they are shorter and the individual notes are simpler. By no means all birds produce song in the technical sense, and even in those species that do sing, the underlying neurobiology related to the production of calls and of song is different. Some well-studied neural circuitry that is central to song production and perception is not equally implicated in calls, as far as we know, and is simply lacking in birds that do not produce song. In contrast to song, calls seem to be dealt with by more general vocal-auditory mechanisms.

Calls and song differ also in their ontogeny. Birdsong is often learned, following slightly different paths in different species. There is absolutely no evidence that experiential learning is necessary for the development of bird calls, however. Birds that have been isolated from conspecifics, deafened, cross-fostered with birds of another species, and raised in other abnormal ways still call normally and appropriately without exposure to models of how and when birds of their own species are "supposed to" call.

Although the calls themselves are not learned, evidence exists in some species for limited variation between individual birds from one region to another; we might describe it as constituting local "dialects." Rather than reflecting direct learning, this variation seems to involve the possibility that a bird's calling can become attuned to a specific model from within a very limited range.

Apart from these examples, development does not involve other interaction with models. A bird does not have to have heard another bird make a call in order to produce it. The calls are evidently determined by the animal's genetic organization, a likelihood that is supported by the fact that some hybridized birds have calls different from either parent. Species that have songs have calls as well, though the reverse is not true.

To summarize, so far we have assumed that the distinction between calls and songs involves some combination of (a) complexity, (b) function, (c) underlying neurobiology, and (d) origin (learned versus completely innate). Calls coordinate the behavior of a pair, a group, or a flock of birds. They may also serve in "maintenance" activities: foraging, gathering together, and responding as a group to outside predators. Calls occur in connection with aggressive interactions, courtship and copulation, and feeding. They may express alarm or distress, and help the birds to maintain contact with one another in a group.

House sparrows (*Passer domesticus*), who may call when they have found food, constitute a more complex example of calling behavior. The intensity of their calling is related to the surrounding circumstances. They call more when (a) the food is more abundant, (b) no other birds are present yet, and (c) the food is easy to share, for instance a number of small pieces of bread as opposed to one large chunk. They call less when there is less food, when others are already there, and when the food does not easily divide. It is easy to see in this behavior an invitation to others to join in the feeding, guided by a combination of generosity and enlightened self-interest. Such variations in the caller's message do not, however, transcend the limitation that there is only one subject for the birds to "talk about." Various subtleties are involved in determining just what the call "means," but there is little communicative flexibility in the system.

Birds of another species, white-crowned sparrows (*Zonotrichia leucophrys*), call to indicate predators. This activity gets the other sparrows to flock around and "mob" the predator, in an effort to discourage it from attacking nestlings. Evidence indicates that these calls too are context sensitive, since the amount of calling depends on the kind of predator and the age of the young to be protected. We are still not seeing a real communicative *system*, merely a single signal whose intensity can vary as a function of the caller's internal state of alarm.

The call of the black-capped chickadee (*Parus atricapillus*) is about as complicated as a vocal signal can get. In aggressive interactions with other chickadees, this bird uses a call that consists of four different sorts of notes. Let us call them A, B, C, and D. A given call has the form $A^*B^*C^*D^*$ —that is, some number of A notes, followed by some number (possibly zero) of B notes, and so on. The combination that occurs on any given occasion reflects the circumstances of the interbird competition. Roughly, the more instances of one note type (A), the more likely the chickadee is to advance; the more Ds, the more likely he is to retreat. A number of nuanced different messages are possible, but the subject matter really cannot be varied.

Black-capped chickadees also give a call referred to as "high-zee" when predators are spotted. From field observation we know that the actual frequency and duration of these calls varies substantially; in fact, different predators elicit calls that can be differentiated to a limited extent. It is possible to relate these differences to a single scale: the more dangerous the situation, the higher the frequency and the longer the call. A continuous dimension results, similar to that found in the bee dances. However, no discrete, distinctively different "words" refer to different predator types (as opposed to what we will find in some other species when we return to this issue in Chapter 7).

A rather interesting issue arises with the behavior of some birds in mixed-species flocks in the tropical rain forest, reminiscent of the broken-wing display of the piping plover. In these groups, some birds keep an eye out for predators and give an alarm when one is spotted, while others forage. Sometimes when food has been found and a large number of different birds are all flocking around trying to get their share, the alarm-calling birds give alarm calls that are spurious (in the sense that no actual predator is around). The others look up or even fly away, and the alarmists have an opportunity to get at the food. These deceptive alarm calls are no different (and do not elicit different responses) from real ones, and we do not know much about the context in which the behavior occurs.

As with the killdeer (*Charadrius vociferus*) or the piping plover, the question arises of whether we ought to say that the alarm-giving birds are "lying." We saw in Chapter 3 that the plover's behavior has a more conservative interpretation: while the broken-wing display is not simply a reflex, we have no basis for believing that the bird is lying so much as performing some behavior that she knows will attract a predator away from her nest. Similarly, the birds in the rain forest flock that give (what we interpret as) a misleading alarm do so because they have learned that the calls will result in a free path to the food. There is no reason to believe the calling birds have any particular sense that the other birds "believe" anything, and thus, that it might be possible to "mislead" them.

Bird calls have a fairly simple internal structure and convey expressive meaning (the bird's own internal state). There is, however, no sense in which they necessarily refer to something outside, as some have argued that the honeybee's dance does. We cannot really think of the birds as "intending" anything by these calls: calling in such-and-such a way is just what the birds do when in a particular state. The state may involve a somewhat complicated interaction with the perceived environment, but if calling has any similarity at all to human communicative behavior, it seems to be at most comparable to human paralanguage.

Birdsong

"But, Father," I said, "it's all right if you get bought right away, isn't it?"

"Yes, but you seldom are, if you're a hen," he said. "People don't often come to an animal shop to buy hen canaries."

"Why?" I asked again.

"Because they don't sing," said he.

You notice he said "don't sing," not "can't sing."

. . .

"Father, I think that's ridiculous," I said. "You know very well that hens are born with just as good voices as cocks. But merely because it isn't considered proper for them to sing they have to let their voices spoil for want of practise when they're young. I think it's a crying shame."

—Doctor Dolittle's Caravan

If the calls of birds are of limited interest regarding their resemblance to any of the properties of human language, what of the other class of avian vocalizations, song? So far I have let that notion remain rather vague, and it is time to get a bit more specific.

Song may range from a simple series of a few more or less identical notes through long arias that may last ten seconds or more. The difference between songs and calls is only partially categorial. Peter Marler, one of the foremost figures in birdsong research, surveys the critical differences. Songs, he says, are "especially loud, longer in duration than calls, often highly patterned, with a variety of acoustically distinct notes." They are "often a male prerogative, with many functions, the most obvious of which are signaling occupation of a territory and maintenance of sexual bonds. Songs are sometimes seasonal and sometimes given year round." Given the importance of song in the reproductive life of the bird, it makes sense that song production is influenced by sex hormones, and song perception may itself affect the production of these substances.

The songs are distinctive from one species to another, of course, as anyone interested in identifying birds knows. Even fairly closely related species can have rather different songs. The typical song sparrow (*Melospiza melodia*) song in Figure 6.1 is much more complex internally than, and con-

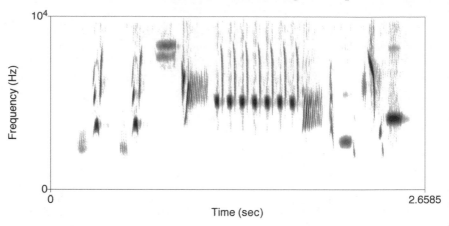

Figure **6.1** Song of the song sparrow

structed quite differently from, the swamp sparrow (*Melospiza georgiana*) song in Figure 6.2.

Comparing these two, we notice that the songs themselves are made up of a number of separate notes, of different types. They occur in a specific pattern, which is characteristic of a given song across repetitions. The pattern of one song may vary from that of another in terms of overall song length, the intervals between notes, the structure of the individual notes, and their arrangement in a particular sequence.

This structure matters: female song sparrows prefer songs that (a) are composed of "song sparrow" notes, and (b) follow "song sparrow" patterns; female swamp sparrows have corresponding preferences for the typical productions of their species. Experiments show that female receptiveness is sensitive to both factors. We can synthesize artificial song sparrow songs made up of of swamp sparrow notes, or of song sparrow notes arranged à la swamp sparrow; when we play these for potentially receptive females of either species, we get some response—more than for song purely of the "wrong" sort, but less than for song that is completely species appropriate.

The songs thus have significant internal structure, a sort of syntactic organization (in the most basic sense), similar to that of the components of the honeybee dance. The component pieces are distinct and get arranged according to some rule—a more elaborate rule than that of the bee dance, though still not very complex.

In fact, the same bird will typically have a repertoire of different songs, songs that are generally similar but distinct from one another. The size

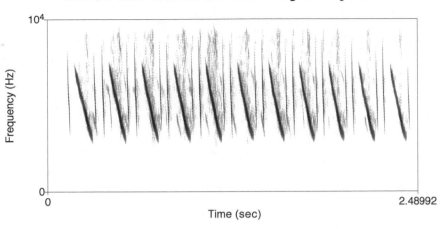

Figure **6.2** Song of the swamp sparrow

of this repertoire varies considerably across species. The common yellow-throat (*Geothlypis trichas*), for instance, sings only a single song, whereas the Carolina wren (*Thryothorus ludovicianus*) chooses from among several (two to ten). The marsh wren (*Cistothorus palustris*) learns several hundred distinguishable songs, including some that he does not sing himself but recognizes in his near neighbors. The northern mockingbird (*Mimus polyglottos*) has an unlimited capacity to learn new songs.

Variety in the song of a single bird serves a purpose, it appears, with respect to the preferences of females. In some species, such as the zebra finch (*Taeniopygia guttata*) and the brown-headed cowbird (*Molothrus ater*), females prefer males who sing more complicated songs or who have a larger repertoire. These properties perhaps reflect more elaborate brain structure, presumably a desirable characteristic to pass on to one's offspring. Apart from advertising his fitness in this way, a male bird may sing a variety of songs so as to make his song comparable to that of his neighbors, or simply for the sake of variety.

In all but the most extreme cases, one might interpret this procedure to mean that the bird has a "grammar" of song, and produces various messages all conforming to the principles of this grammar—much as an English speaker produces various messages all conforming to the grammar of English. In one sense, it is an apt comparison, because the bird's repertoire is based on a notion of song that is not limited to a single pattern but is still narrowly constrained. On the other hand, it would be a mistake to treat the song variants as different messages, since the import is exactly the same regardless of which form is employed. Rather than different messages, what we have is a constrained range of variation in the way the *same* message can be realized.

The most basic difference among songs in terms of biological importance is that between the songs of one species and those of another. A baby bird comes into the world with a genetic predisposition to develop the songs of his own species, regardless of the complex range of acoustic events he may hear in the surrounding environment. Song sparrows raised among swamp sparrows, for instance, will still generally sing like song sparrows. This effect is not absolute: in some experiments birds exposed primarily to the song of another species at an appropriate stage of development did in fact acquire that song rather than one appropriate to their own species. The limits on the possibility of learning heterospecific song are fairly narrow and result from the similarities between song types and the lack of more appropriate models. Given a choice of models, birds will pick out and learn the song of their own species.

Songs often—perhaps even usually—display differences of dialect from one geographic region to another, within a single species. That is, local variations typically characterize the song, falling within the broad range of possibilities available to the species in question. These differences (unlike the more basic ones between the songs of distinct species) are not genetically determined. If we move a baby bird into a different area, the song he develops will be appropriate to the local dialect.

In part, it is history that determines dialect. Where a species has a wide repertoire of possible songs, local variation may merely reflect the fact that only some models and not others are available in a given region. This state of affairs will of course persist to the extent that later generations learn from (the descendants of) those same models.

In part also, dialect formation may reflect social adaptation. In some species, young males who sing like their neighbors are more successful in attracting females, which means that conformity to the local model has very direct consequences for reproductive success. Since females often make their choices based on properties of the male's song, these preferences will tend to shape the songs found in a given area.

Even in the majority of temperate climate songbirds, where females do not sing, they still prefer songs selected from within the locally appropriate range of variation. Females apparently do some song learning too, despite the fact that they do not themselves sing the songs that they learn. The neurological structures underlying song production are much less prominent in such females, but are not altogether absent.

While much is fascinating about the songs themselves and their structure, we must leave that to field guides and studies specifically devoted to that topic. Here we address two topics in the study of birdsong that have a broad impact on the study of communication: first, the physiology and neurophysiology underlying the mechanisms of song production and perception, and second, the path by which control of song develops in the baby bird. Both areas contain striking parallels to facts about human language.

Song Production and Perception

Let us return to the song of the song sparrow. As with frog croaks and human speech, we could represent the song directly as a pattern of pressure variation in the air, with an oscillogram. It is much more informative, however, to approach it from the point of view of a spectrogram as in Figure 6.1, which displays the distribution of energy across a spectrum of frequencies as a function of time.

What did the bird do to produce this song? The analysis of vocal signals is most developed, naturally enough, in connection with human speech. As we saw in Chapter 5, speech is generally analyzed in terms of what is usually called a source/filter theory. The sound that emerges in speaking is the product of two factors: an energy source and the filtering properties of the vocal tract.

When we humans talk, we produce airflow from the lungs, and the vocal folds open and close rapidly to produce a sequence of pulses. The train of puffs of air through the larynx constitutes an energy source, but one that is common across a range of sounds, so it cannot be the only element involved in differentiating one sound from another.

These differences emerge when we move our articulators around to produce a particular filtering effect. Some of the acoustic energy introduced into the vocal tract comes through, whereas energy at other frequencies is attenuated or absorbed. What distinguishes one vowel from another is the difference in this pattern of filtering. We distinguish a vowel's *quality* (for example, [i] versus [a]) from its *pitch*. The former results from variation in shape of the filtering vocal tract, while the latter is a product of variation in the nature of the energy source.

This much is already familiar in the case of human speech, but does the

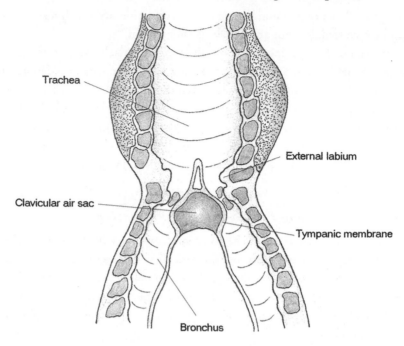

Figure **6.3** Structure of the avian syrinx

same kind of analysis apply to birds? We immediately note one significant difference. In oscine songbirds, the larynx is used only to open and close the air passage globally, not to create vibration. Laryngeal vibration thus does not provide a source of sound energy, which is instead the responsibility of the *syrinx* (Figure 6.3).

The syrinx is an organ with two separate sides, and it produces acoustic energy in two ways. First, the labia (at the opening on each side) can be positioned to vibrate, thereby producing a sequence of pulses similar to human laryngeal vibration. Second, in other positions the labia produce controlled turbulence at the outlet from (each side of) the syrinx. Such turbulence is another form of acoustic energy; basically, this sound source is similar to that involved in whistling. Energy from turbulence (rather than from a controlled sequence of puffs) also plays a role in human speech, in the production of consonants such as [s] and [t].

A second difference between birds and humans is that the bird's suprasyringeal vocal tract is extremely short and not very controllable. It has a filtering effect, but one that in most species is more or less a constant. Thus, control of the output in song is almost entirely a matter of managing the syrinx. The bird is not at a total loss, however: as we know from whistling, quite a bit of variation can be produced in this way.

The two sides of the syrinx are more or less independently controllable in songbirds, so the result can be rather complex. In several bird species, one side of the syrinx (typically the left) is larger than the other, and so produces lower-frequency notes. The bird's song takes advantage of this fact, allocating some notes (or parts of syllables) to one side of the syrinx and some to the other—with both voices cooperating at times. The result is a complex pattern of vibration in the air, which we hear as birdsong.

In other birds, the independence of the two sides of the syrinx is employed to help out with breathing. The bird produces a note with one side, keeping the other side closed. He then opens the other side briefly and takes a minibreath, closes it and produces another note with the first side, and so on. The song therefore goes on for much longer than might be possible if it were limited by the breath cycle of the bird's tiny lungs.

What happens on the receiving end? That is, how does this pattern of vibration in the air give rise to sensation in the other birds that constitute the singer's audience? Again, we start with the human ear. As we have already seen, vibration in the air is transmitted via the middle-ear structure to the cochlea. The cochlea is a fluid-filled tube where vibrations cause a membrane in the middle of the fluid channel to move. These movements of the basilar membrane stimulate the hair cells, which produce nerve impulses along the auditory nerve. Hair cells positioned at different points along the basilar membrane are sensitive to different frequencies and thus supply the auditory nerve with what is in effect the same decomposition of the signal as seen in a spectrogram.

At this level of generality, birds are pretty much like people. Their ear is constructed with a cochlea that responds up through about 12 kHz and that provides sensory input to the brain by essentially the same mechanisms as in human audition.

So, one asks, what does the brain do with this input, and what is its role in controlling the performance of a singer? We can supply a much fuller answer for birds than for humans. Jokes about "birdbrains" aside, the brain of the bird has a level of complexity more manageable than our own, and in addition we allow ourselves to perform experiments in vivo on birds that we would never contemplate on humans. As a result, we have a rather satisfactory map of the areas of the brain that are involved in producing and perceiving song.

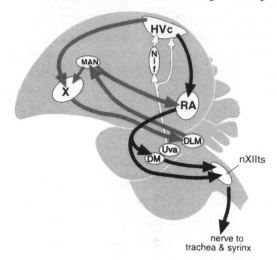

Figure **6.4** Brain regions involved in birdsong and their interconnections

Figure 6.4 sketches the main areas of the bird's brain that are relevant to our questions. The control of song is centered in connections among several of these areas, especially those conventionally designated *HVc, MAN,* and *RA.* Without delving into the relevant neurophysiology, we note that in songbirds four functions are subserved by this apparatus: (a) initiating nonsong calls, (b) initiating and controlling song, (c) perceiving song, and (d) learning song.

One of the major findings in human neurophysiology is that of *hemispheric asymmetry.* Although the two sides of the brain are roughly similar anatomically, they are specialized in somewhat different ways for particular functions. For instance, most structural language functions are concentrated (in the great majority of adult right-handed males) in the left hemisphere, while nonstructural, emotional, and paralinguistic functions are mostly in the right hemisphere.

The scope of this observation is sometimes exaggerated. It appears that the degree of asymmetry between hemispheres, both anatomically and functionally, is less in women as a population than in men; left-handed individuals may be more nearly symmetric than right-handers, with a small proportion of left-handers having structural language functions lateralized in the right hemisphere instead; and overall, the degree of lateralization varies somewhat among individuals. Despite the "noisiness" of the data, hemispheric asymmetry is a distinct tendency in humans with respect to language functions. Since birds have two hemispheres too, we might ask whether their brains are lateralized as well with respect to song.

The answer is yes and no. Lesioning just one side of the bird's brain has different overall effects on the song depending on the hemisphere affected. In canaries, for example, lesioning the left HVc results in the bird's sitting up on his perch, fluffing out his feathers, and opening his beak as if to sing, but no sound comes out. Actually, this effect is somewhat puzzling; although birds have two distinct hemispheres in the brain, there are few anatomical distinctions between them. In addition, bird brains do not have a *corpus callosum* (the structure in the human brain that connects the two hemispheres). The two avian hemispheres are largely independent, although a few connections exist.

The song-system motor pathways of birds are uncrossed, so the left hemisphere controls motor activity on the left side, and the right hemisphere controls activity on the right. Since the bird's syrinx has two parts (Figure 6.3), we might think the left syrinx is controlled by the left brain hemisphere and the right syrinx by the right. In fact, though, physical coupling between the two sides of the syrinx increases the complexity of the matter. What the bird does with the muscles on one side affects the configuration of the other as well, so it seems that different, but interrelated, parts of the song can each be controlled separately by the two hemispheres. In addition, when the nerve controlling the left side of the syrinx is cut, the bird cannot sing, even though he can breathe normally. Cutting the nerve to the right side of the syrinx has almost no effect on either breathing or song. Some crucial aspect of gestural coordination in the song must be contributed by the left side, although it has effects on both sides. Clearly a kind of lateralization, it is apparently not the same kind we find in humans.

It is particularly interesting to note that the motor control areas involved in song production are *also* involved in song perception. Heather Williams and Fernando Nottebohm have shown (in the zebra finch) that primary motor areas involved in song production, right down to the nucleus innervating the syrinx, respond when the bird is hearing something with roughly the auditory properties of its own species' song—but there is much less response in these areas when the bird hears stimuli with auditory properties that could not be song, or even when it hears its own song backward.

The attentive reader will note immediately that this discussion offers strong support for the motor theory of speech perception, at least for birds, since the system whose basic function is to control production is also actively involved in perception. I do not intend to imply that the bird is literally singing along with what it hears (though birds probably come closer than people to doing so): I just mean that the production system is involved, and hence that its dimensions of control can be expected to play a part in perception.

Other results from neurological investigations concern the mechanisms involved in learning. For example, a comparison of species in which song is learned (such as zebra finches) with those in which it develops spontaneously without need for a model (such as eastern phoebes, *Sayornis phoebe*) shows that brain areas that are essential to song production and learning in the former (for instance, HVc) are simply absent in the latter.

Birds whose song is learned have not one but two neural pathways that are involved in song production. One of these seems to be specially involved in learning, because it becomes redundant once song is fully developed (crystallized). Thus, specialized brain physiology is intimately connected with the learning process, which is of course related to the notion that the *process* of song learning is innate even though the specific song to be learned is not.

But we are getting ahead of ourselves. It is time to look more closely at how baby birds come to sing.

The Ontogeny of Song

"As you may have observed," Pipinella went on, "young birds talk and peep and chirp almost immediately they are out of the egg. That is one of the big differences between bird children and human children: you see before you talk, and we talk before we see."

"Huh!" Gub-Gub put in. "Your conversation can't have much sense to it then. What on earth can you have to talk about if you haven't seen anything yet?"

"That," said the canary, turning upon Gub-Gub with rather a haughty manner, "is perhaps another important difference between bird babies and pig babies: we are born with a certain amount of sense, while pigs, from what I have observed, never get very much, even when they are grown up. No, this blind period with small birds is a very important thing in their education and development."

—*Doctor Dolittle's Caravan*

In discussing the vocal signals of birds and other animals, I have distinguished between signals that are innate and those that are learned. What exactly do we have in mind when we say that a call, song, or other signal is learned? Essentially, we mean that the signal in question has the form and usage that we observe in an adult animal on the basis of some necessary interaction with the environment over the course of its development. That is, the animal uses the data of experience, in conjunction with its own biologically determined constitution, to arrive at the shape and conditions of use of a "learned" signal. Innate signals, in contrast, arise in essentially their final form on the basis of the animal's genetic background alone. Of course, an innate signal may vary somewhat over the animal's lifetime if simply as a consequence of maturation. For instance, frequencies may change as an animal grows bigger, but for reasons that are independent of what the animal finds in its surroundings (such as corresponding signals by others of its species).

Saying that learning involves an interaction with the environment, though, does not pin matters down very precisely. Vocal signals can depend on experience in a variety of ways. For instance, vervet monkeys have alarm calls that they give in the presence of certain types of predators. The calls themselves appear to be largely innate (though aspects of learning may be involved in some of the noncentral cases). That is, young vervets produce the correct calls at a very early age with no need to hear others produce them. Still, experience is relevant, because baby vervets use the calls in a much wider variety of circumstances than adults. As they mature, the behavior of other vervets provides information that helps the animals refine the conditions under which, say, an eagle alarm call is appropriate, though the nature of the call itself does not change.

Experience might likewise be relevant to the development of dialects in the calls of a variety of species, including birds such as petrels, prairie dogs, and several other birds and mammals. The association of specific forms of a call with a particular group or region might be genetic in origin. That is, the local population might have developed largely in isolation from others of the same species, and small variations in calls that arose have been passed on to offspring. This kind of dialect formation is independent of data from the environment.

Dialects could develop in another way, too. Suppose that a given species has a particular call with a range of possible variations. Any individual of the species may produce and respond to any of these variations, a capacity that is genetically determined. Now suppose that some individuals within a given group or geographic region make use of only a limited subset of these possibilities. It might then be the case that others in the same group would similarly restrict their calling behavior. And if different subsets of the possible calls came to be standardized in this way in different places, we would have local dialects. Such dialects would not be *shaped* by the animal's experience, only *selected for.*

Yet another kind of interaction with the environment is illustrated by the calls of some species of bat. In the lesser spear-nosed bat (*Phyllostomus discolor*), mothers and their pups use a contact call to identify one another, when the mother returns to her pup with food or when the pup has become lost. Individual bats have distinct calls, and they are reasonably proficient at distinguishing individuals from one another on this basis. The baby bats produce a version of this call early in life, without apparent basis in experience; but as they develop, their calls fairly quickly come to resemble those of their mothers. The effect of experience is much more direct here, shaping an existing call in particular ways rather than choosing one from a range of possibilities.

Among these bats, it is not only infants in whom experience shapes their innately based calls. Adult bats forage together, and they maintain contact with one another through a distinct "screech" vocalization that varies from one group to another. The bats discriminate between screech calls given by members of their group and by members of other groups, though experiments have shown that they are unable to distinguish specific individuals on the basis of differences in these calls. Because the members of a given group are not usually close relatives, a genetic basis for the similarities within a group is unlikely. When an adult bat joins a new group, it takes him about five months to assimilate his screech calls to those of the other members of the group—another instance in which experience plays a shaping role, but this time somewhat later in life.

This kind of selective or shaping influence is interesting and important, but I distinguish it from learning in a stricter sense. By the latter I mean a case in which the specific form of the vocalization is not innately present at all. Instead, the organism comes into the world with a kind of schematic knowledge, a template or a set of parameters, that defines a range of possibilities. Early in life, experience provides models (via the behavior of other members of the species) that fall within the range of this system, and the animal develops a pattern of vocalization based both on the specific models and on the properties of the general pattern.

Let me call this (true) *vocal learning,* as opposed to the kinds of social modification discussed above. It is of particular interest because it seems to be the way human knowledge of language develops. The only other animals that appear to learn vocalizations in this way are birds, and then only a limited subset of them.

It is in only three of the twenty-seven orders of birds that song is learned in the sense I specify above. These are songbirds in the narrower sense (oscines, suborder *Passeri*), parrots and their relatives (family Psittacidae), and hummingbirds (family Trochilidae). The song system in each seems to have evolved independently of the others. Peter Marler observes that "oscines [have] uniquely complex vocal apparatus, and [a] specialized network of brain nuclei that constitutes the 'song system,' lacking from the brains of suboscines, and as far as we know, from the brains of all birds with *innate* songs."

We now know that this neuroanatomical specialization is not limited to the oscines. Studies of the nervous systems of other species that learn their songs have uncovered brain nuclei that are shared among songbirds, hummingbirds, and parrots—the groups that have developed song based on vocal learning. Remarkably, the relevant structures, which have evolved in the necessary configuration independently in the three groups of birds, are absent in other, nonlearning species. "This would suggest that the evolution of these structures is under strong epigenetic constraints."

Despite these fascinating possibilities, not much is known about the specifics of song learning in hummingbirds and parrots, and our discussion here is limited to the well-studied cases of oscine songbirds. Marler's pioneering studies first drew attention to the striking parallels between song learning in birds and the development of language in human infants.

In other birds, the song is completely innate. Young phoebes, for instance, sing nearly correctly at the age of about two weeks, and the course of improvement depends solely on better control of the vocal apparatus as the birds get older. Crucially, they will sing correctly if (a) raised in isolation from other phoebes, or (b) deafened shortly after hatching. In contrast, birds that learn need experience and models, else they will produce only pathetic squawks instead of real song.

Just what is learned varies considerably: "Some learned birdsongs are relatively simple, on a par with those that are *innate.* Other *learned* songs are extraordinarily complex, with individual repertoires numbering in the tens, hundreds, and in a few cases, even in the thousands."

The role of learning, too, is diverse across species. In the cuckoo (*Cuculus* species), even a complicated song is unambiguously innate, since cuckoos can sing correctly without having heard the songs of others. The trait is adaptive for this bird, for cuckoos typically lay their eggs in other birds' nests. As a result, the babies may not have other cuckoos nearby to serve as models for song learning.

In other species, though, learning is definitely involved. Its nature can range from simply identifying some individual conspecific's song and copying it, to acquiring an appropriate pattern from within a limited range of possible choices (chaffinch, *Fringilla coelebs*), to relatively free learning (bullfinch, *Pyrrhula pyrrhula*).

Birds with a fixed repertoire learn their songs during a critical (or "sensitive") period in early life. It usually lasts less than a year, and may be shorter. In the marsh wren, for instance, the sensitive period is only about two months long. Yet it is common for this phase to extend into the first breeding season after the bird is born, ensuring that young birds have the opportunity to learn from their experienced neighbors. Information about possible song models acquired during the sensitive period is not generally put to immediate use, but is stored in memory and guides later stages of the learning process.

In accordance with Marler's research, the course of learning is seen in terms of four distinguishable phases. The times given here are for white-crowned sparrows, the birds Marler first studied in detail. Other birds vary the schedule somewhat, but pass through the same general stages.

Learning begins with an initial silent period which may last up to eight months in some species, when the juvenile bird does not sing at all. During this stage memorized song components are stored in the brain. In the white-crowned sparrow, this period ranges over the first 15–35 days. Although the bird does not sing at this stage, other vocalizations (unrelated to song) such as a "begging" call may occur. The young bird is listening to the song of conspecific models. These models are heard, internalized, and saved for later use in shaping the bird's own eventual song.

Next, during a period roughly 25 to 40 days after birth, the bird begins to produce what is known in the literature as *subsong*. It consists of "relative soft, broad band, unstructured sounds. This stage is thought to be the process by which a young bird calibrates his vocal instrument, akin to babbling in human infants." Within a few weeks, the sounds produced begin to resemble those found in songs typical of the species.

From about 35 to 80 days is the period of *plastic song*, "marked by the gradual approximation of the young male's song output to the stored model(s)." At this point the young bird's song is recognizable as falling within the range of his species, though he is still working on it. Finally, at about 90 days, comes *crystallization*, the stage at which experimentation seems to end and the bird's overall song becomes fixed. Full song is not necessarily produced at this point, though. In white-crowned sparrows full song does not appear until the following summer, at around 240 days.

In some species, many different songs are sung during the plastic song period and all but a limited number are discarded at crystallization. What determines which songs get dropped? In birds raised in a laboratory, what happens is more or less random. But in nature, the selection may be affected either by competition with other males (it helps to be able to sing something like your rival's song) or by reinforcement from females.

An interesting variation on this last possibility is provided by the cowbird (*Molothrus ater*). These birds are often raised in other birds' nests, so (like the cuckoo) they do not generally have adequate conspecific models in infancy. What they produce as they enter maturity tends to contain elements from other birds' songs that they have heard, plus elements that are more "cowbird-ish," but not real cowbird song.

When these birds go out to compete for mates, they sing this mishmash and watch the female cowbirds. When they hit notes that are suitable, the female waves her wing in a particular way—a kind of "flirting." The male picks up on that encouragement and includes more of those elements in his song. Eventually, he comes to sing in a completely appropriate way. Here it is not a model (in the form of another adult male cowbird) that shapes his song, but rather the female's preferences—which must themselves be innate, in that female cowbirds are no better off than males as far as the availability of adult song models is concerned. This kind of final shaping goes on somewhat later than in ordinary song crystallization.

If we set aside for now the sort of fine-tuning that gives specific form to the cowbird's song, much of what we see in birds is strikingly parallel to human language acquisition—given the fact that the bird has only to learn to speak, not to connect the details of speech with the details of a vast range of possible messages. What has to develop is something akin to the phonology of a human language, but nothing that corresponds to its syntax.

The full range of ways in which the nature of the bird's learning process determines its outcome varies considerably, and in some cases this process must be much more specific than it is in humans. A human baby can acquire with equal facility the system of *any* natural language that happens to be spoken in the surrounding community. This flexibility represents only one end of the scale among birds, though, many of whom apparently ship from the egg preset to acquire exactly one specific song (or one from a tiny set). They will learn exactly this song (or small set of songs), but still require empirical experience to reinforce the connections that crystallize exactly the right one.

Despite variation of this sort, song learning in birds and language learning in humans is fairly similar. Each comes into the world with the ability to learn patterns from within a particular range. Each makes use of models provided by early experience to determine a particular instance from within that range (a set of songs, or a particular language). Birds and babies are both capable in principle of learning any of the patterns typical for their species, though the number of such patterns that are made available by the species' biology varies widely.

As we have seen, song learning takes place during a specific period in the young bird's development. The song of birds that are isolated from adult song models during this time will later show deficiencies, although the song will typically contain at least some elements of "proper" song. To return to the song sparrow, the sensitive period during which song models are learned is roughly between days 20 and 70 after birth. Birds that are only exposed to adult song before this (between days 3 and 7) and are otherwise isolated produce abnormal songs that are closer to subsong than to those of other adults. Birds whose only opportunity to hear adult models comes late in the sensitive period, between days 50 and 71, show some deterioration in their songs as adults; while birds that do not hear adult song until much later, after 300 days or so, again do not get much beyond the stage of subsong. In contrast, birds whose only opportunity to hear appropriate models comes between days 35 and 56 nonetheless sing normally as adults.

Birds that are deafened, or deprived of normal sensorimotor feedback during the sensitive period, produce even cruder approximations to normal song as adults. Feedback plays an essential role in the learning process. With respect to both the role of song models at the right time and the importance of feedback, we can contrast learned song with innate calls, which require neither experience nor sensorimotor feedback to develop normally.

The way song development proceeds based on models heard during the sensitive period is illustrated in elegant detail in work conducted by Dietmar Todt and Henrike Hultsch on the common or European nightingale (*Luscinia megarhyncos*). Nightingales learn several hundred songs, and they sing a great deal. They are excellent mimics of the songs of other nightingales. Furthermore, they listen to (and learn from) songs presented over loudspeakers, which makes them ideal for study as opposed to birds that only treat songs as "real" when they can see the singer.

The nightingale's many songs are built up from a basic repertoire of about forty distinct notes. Each song is acquired on the basis of only a few exposures, a process presumably made easier by the fact that all of them conform to a limited number of patterns. Todt and Hultsch divided these patterns into four distinct parts, which we refer to as A, B, C, and D. They then manipulated real nightingale songs by removing or repeating the song parts, and played the results to song learners.

Table **6.1** Results of manipulating song elements in nightingale learners

Modification	Acquisition	Results
Omit section A	Normal	Novel A sections invented
Repeat section B	None	—
Triple section C	Normal	C reduced to normal
Repeat section D	Normal	(Plastic song) D repeated
		(Full song) D corrected
Concatenate two songs	Normal	(Plastic song) sung together
		(Full song) split apart

In their data, we can see ways in which the bird takes in a model song and builds a new song on the basis of what he has heard, a song that conforms to the properties of nightingale song even when the model did not. In some cases (such as an extra B section), the model is not treated as proper song at all and does not lead to learning. In others, violations of the "grammar" of nightingale song are repaired.

The stages of song learning, the role of a critical (or sensitive) period during which input has to be available for song to develop normally, and the creative role of the learner are all comparable to what we find in human infants. Obviously, a given bird has a specific range of systems that it is capable of learning. The specification of that range is like what linguists refer to with respect to language as the role played by Universal Grammar. Just as the child does not have to invent the idea of language anew on the basis of observation, the bird does not have to discover how song works. He has only to learn which song to sing, within some constrained range of possibilities.

In cases similar to that of the nightingale, we observe that at least some birds, during the plastic-song phase, produce song elements that are not the same as any they have actually heard—but still fall within the range of possible song elements for their species. The "errors" children make during the course of language learning virtually always correspond to things that could occur within some possible grammatical system—just not the one they will eventually settle on as appropriate for their community. The child (or the bird) is still working out exactly what system the language (or the song) to be learned instantiates, from among a determinate set of possibilities.

In birds, some of these errors may persist into adult song as creative innovations, which can then spread within the community of the species. Analogy with the kind of historical development that linguists call *sound change* does not seem far-fetched, although it would require more substantiation than we can supply here.

The developmental sequence we observe in baby birds learning to sing shows us how a particular kind of learning takes place. It seems incontrovertible that this learning sequence is determined by the bird's biology: vary the species, and the range of systems that can be acquired changes, regardless of the nature of the actual sound input. Song sparrows cannot learn to be bluebirds (*Sialia sialis*) even if bluebird song is all they hear (although they *can* learn to sing somewhat like swamp sparrows, a closely related species, when that is the only model available). If the parallels with human language are as convincing as they seem, that suggests further reasons for taking human language to be biologically determined too, even though the range (and internal complexity) of the systems involved seems much greater.

It is well known that regional dialects develop within songbird populations. These are often remarkably specific and quick to emerge. Evidently, their basis is the role of the internalized song model in shaping the eventual song, within a range of possibilities appropriate to a species. The central role of the learning process in the development of such variation implies that regional dialects should never arise in birds for whom song is innate and not learned.

In fact, that prediction turns out to be wrong. Just such regional dialect formation has been reported in at least one group of (nonoscine) birds, the petrels (family Procellariidae). Virtually all of the literature on song, its origins, and related matters deals with songbirds, in whom the song is learned. Little significant work has been done with species where it is known (or at least assumed) that the song is innate. In the case of petrels, it has been demonstrated that the repertoire of vocalizations is innate, since it develops equally well in normal birds, cross-fostered and isolated birds, and deafened birds. It is also true, however, that significant variation exists between populations as close as two kilometers from each other. How can this be, given the account we have developed above?

The path to such variation is by processes of social modification. We note first that the calls and songs of any animal species are subject to a certain amount of interindividual variation. Animals are not all physically alike, however difficult they may be for members of another species, such as ours, to distinguish. And this variation undoubtedly can be "sculpted" in various ways.

Consider a vocalization that serves in some species as a mating call. Let us assume that the production of this call is entirely innate, but still subject to some interindividual variation in its form. We assume further that female preferences in response to this call, while again entirely innate and independent of experience, are nonetheless subject to minor interindividual variation. Under these circumstances, within a small, somewhat isolated population these preferences could provide evolutionary pressures favoring some call variants and disfavoring others. That is all it would take to produce regional dialects, even though the process might take a bit longer than when aided by a more flexible learning process.

Even without this kind of selective pressure, we can imagine that within a group of birds showing limited (but nonnegligible) variation, regionally distinctive preferences could develop. This process would not be much different than if I (as an adult) move from Guilford, Connecticut, to Gilford, England. My speech is likely to change somewhat over time, though I may never sound entirely like a native. Even a system that is fully complete can change somewhat—not in its basic structure, but in favoring some possible low-level manifestations over others to accommodate to surrounding speech.

We can distinguish this kind of shaping from the learning we see both in songbirds and in humans. In cases like that of the petrel, experience is not necessary for the system to develop normally. The role of experience is simply to take the raw material of natural variation and favor some variants while disfavoring others—either by evolutionary pressure or by local assimilation. In true learning, experience interacts with the range of possibilities allowed by the biologically determined nature of the species. This interaction between "nature" and "nurture" is essential to the process that allows the system to develop at all. It provides human language learners with a kind of flexibility that (at its extremes) is rare (and probably unnecessary) in birds. As a by-product, it also makes the system responsive to local variation.

The ability to produce (and understand) signals appropriate to its species arises along paths that depend heavily on the bird's biology. In most instances (calls in general, song in some species as well) the system is entirely innate, in the sense that it requires no significant interaction with experience to develop fully and completely. In the broader context of cognitive systems, think of the way the olfactory system arises in humans, who need no experience to develop a system distinguishing the basic tastes.

In other bird species, some interaction with experience is necessary, though the possible effects of that experience are restricted because of the limits on what kinds of song can develop in that species. This situation is somewhat like that of the visual system in cats (and, as has now been established, in humans as well). Given a visual environment that is not markedly deficient, the visual system of the animal will develop in a way whose outcome is essentially fixed in advance. If visual stimuli of certain general types are not available during a critical period of synaptic formation during maturation, the right connections do not form and the corresponding aspects of visual ability will be lacking. As in the case of some birds' song, the system needs experience in order to develop, but it develops in a fixed way once it has that experience.

In yet other birds, the role of experience is rather greater in that they can develop a range of songs, some of which may not be exactly like any song they have actually heard. Even here, the role of biology is strong. Again, there is a limitation on the *kind* of song the birds will acquire. The ability to sing develops in a programmed way that requires experience during the appropriate sensitive period. This experience is internalized and then serves to shape later development through the phases of subsong and plastic song through to the stage of crystallization.

No one would seriously doubt that the control of birdsong learning is largely organized as a function of the bird's neuroanatomy, and thus of its biology. In some birds development as a whole is more or less innate, since the bird can come to sing its song with little or no environmental input. In others, the song control system develops on the basis of an interaction with data provided by the environment. The productivity of this interaction is dependent on physical changes in the bird's neuroanatomy, changes that are controlled by its specific genetic program.

Birds that learn to sing are not all alike neurologically. In most species (for instance, the zebra finch), song is learned once, in the first year, and stays constant throughout the rest of life. In other species such as the island canary (*Serinus canaria*), new songs are learned each year.

A theory of what might account for these differences, associated with Fernando Nottebohm, is based on cyclic patterns of nerve cell birth and death that vary from one species to another. When we look at neurons in HVc, one of the song-relevant parts of the canary's brain, "cell birth and death are associated with song learning and song forgetting, respectively." A related but independent observation in the chickadee indicates that the growth of new neurons in the hippocampus (a region important for spatial learning) is greatest during the time of year when the birds store seeds. Thus, neuron growth in general may be associated with occasions on which birds create new memories, whether of songs or of where they have stored food.

When we compare canaries with zebra finches, we find that birds which learn their songs once and for all also add large numbers of neurons in the HVc area only once, in their youth. The correlation between neuron growth and song learning is not perfect, however. Song sparrows, which learn their songs once as babies, show cyclic patterns of neural birth and death in the HVc. Such patterns, then, are not exclusively related to the possibility of learning new songs, but they may nevertheless be relevant to it in species such as the canary. Neurogenesis in HVc definitely plays a role in song, though the exact nature of the connection with learning needs to be clarified. New neurons tend to appear in this region every year in species such as the canary that learn new songs, but not in most species that learn only once. To the extent that this observation holds, it provides a neurological basis for the observation that learning is associated with critical or sensitive periods, and that the timing of these is a consequence of physical changes that play themselves out during a bird's maturation.

Considerable evidence is available for genetic control of song learning in the canary, in particular. First of all, it is possible to breed canaries selectively for the kinds of song they will sing. This fact is prima facie evidence that the range of songs that can be learned is a function of the bird's genetic endowment. When you cross-foster canaries of two sorts, one tending to sing only low notes and the other mostly high, the baby birds will indeed learn from the models they are given — but they learn primarily those aspects of the models' songs that their breeding predisposes them to specialize in. That is, a "mostly low note" bird raised with "mostly high note" models will learn from the songs he hears, but what he learns will be primarily those parts of the song that fall within the range we would expect on the basis of his parent's song. Assuredly, learning is taking place, but learning that is strongly shaped by the organism's biological program.

The Human Language Organ

I have dealt in the preceding section with the production, the perception, and especially the development of song in birds, but I have drawn attention repeatedly to the parallels that can be made with a particular picture of human language learning. On this view, the development of language in human children is similarly driven by human biology. Normal language acquisition in humans, like song learning in birds, takes place preferentially during a specific stage of maturation. Even as what a bird is capable of learning is a function of its species-specific biology, so also what humans are capable of learning as a first language is undoubtedly determined by our genetic program.

The evidence from the zebra finch suggests that perception of the song of conspecifics proceeds in a special way that involves the neurological systems underlying production. These are systems not implicated in the bird's general perception of other auditory stimuli. Similarly, we saw in Chapter 5 that the human perceptual system operates in a highly specialized mode when processing signals that have the property of speech — a mode distinct from the way the auditory system works when processing other signals.

These points all suggest that language has a special status in the brain. I now briefly survey the kinds of connections that can be made between language and the brain, and then flesh out the path by which humans develop the phonological system of their native language. There is much more to human language than this, of course. We will see some of its other aspects in later chapters, but here is where the parallels with avian song learning are most obvious. And we have no reason to doubt that exactly the same kinds of biologically driven mechanisms are at work in developing other aspects of language, aspects that have no substantive parallel in other animals apart from being within the genetically determined range of possibilities open to the species.

The Neural Substrate for Human Language

It is evident that human language, like the communicative capacities of other animals, depends on the specific structure of human beings: our vocal organs and our ears, and most especially our brains. An abundance of evidence makes it clear that our linguistic capacity is based in myriad ways on our genetically determined constitution as members of our species. In the case of other animals, we accept the biological determination of their specific means of communicating, and much else in their behavioral repertoires, as a matter of course. When it comes to ourselves, this suggestion is somehow much more controversial. The case for genetic mechanisms that determine the human language capacity, and more generally for a "language organ" that is part of our makeup in the same way as our kidneys, has been made at length elsewhere and I will not rehearse it again.

We do need to acknowledge that the nature of language must reflect in rather precise ways the organization of particular body parts, most important of which is the brain. That may seem obvious today, but the relation of the brain to capacities such as language has not always been taken so much for granted. Historically, we find mention in Babylonian and Assyrian cuneiform tablets from as early as 2000 B.C. of a connection between disorders of intelligence and brain disease. An Egyptian record from around 1700 B.C. makes a link between an injury to the brain of a particular patient and his suffering a loss of speech. Later, though, the relation of the brain to cognition and language became somewhat more obscure. Aristotle, for instance, taught that the brain was really a large sponge, whose function was to serve as a radiator to cool the blood.

Only with the development of high-powered microscopes did it become possible to see enough of the structure of the brain as an organ to formulate a coherent picture of its organization and functions. The notion that the brain, and the nervous system more generally, is a collection of cells that communicate with one another at specific connection points — the *neuron doctrine* — only dates from the work of Ramón y Cajal and other neuroanatomists of the 1880s and 1890s.

On a larger scale, two distinct views of the brain have been in contention for much of recent history. One has its modern origins in the theories of Franz Gall, who believed that the brain is composed of a large number of faculties, each specialized for a limited range of functions. He offered detailed charts of this kind of organization, similar to that in Figure 6.5. Notice that in this chart, the organ of "Language," area XXIX, is located just below the eye — somewhat peculiar, in light of the absence of brain tissue there.

Gall also held the view that exercising a mental faculty caused the corresponding part of the brain to get bigger, pressing against the skull and causing bulges (or, in the case of underutilized faculties, depressions). That gave rise to the pseudoscience of *phrenology*, which tended to give "faculty psychology" a bad name (from which it has not entirely recovered, even though the physical literalism of Gall's version has been largely abandoned).

At the beginning of the nineteenth century, Marie-Jean-Pierre Flourens performed some experiments on animals in which he excised parts of their brains and then looked to see what specific abilities were affected. By and large, all of his experimental subjects were affected globally. That is, rather than losing just one particular function, they became totally passive. From this he concluded that cognitive capacities are not localized in the brain, but rather are distributed globally, so that injury to any part results in overall degradation. This view that the brain represents a single very general faculty, rather than a number of individual and specific ones, is sometimes called the *aggregate field* view of the brain. Steven Pinker caricatures it as "the brain as meatloaf."

The aggregate field view is congenial to one particular notion of human beings. If specific faculties are in fact anatomically localized, that increases the extent to which we see those faculties as based in the properties of specific material in the physical world. If we think that most higher mental faculties are actually properties of an incorporeal soul, this materialist view

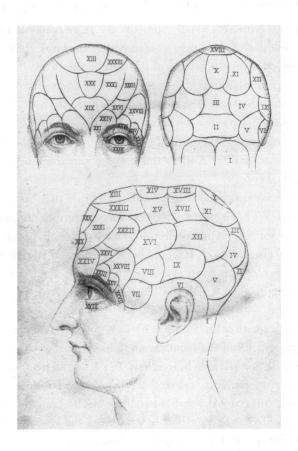

Figure **6.5** Regions identified as personality organs according to the classification of Gall, with additions by his follower Spurzheim: I, Amativeness; II, Philoprogenitiveness; III, Inhabitiveness; IV, Adhesiveness; V, Combativeness; VI, Destructiveness; VII, Constructiveness; VIII, Covetiveness; IX, Secretiveness; X, Self-Love; XI, Approbation; XII, Cautiousness; XIII, Benevolence; XIV, Veneration; XV, Hope; XVI, Ideality; XVII, Conscientiousness; XVIII, Firmness; XIX, Individuality; XX, Form; XXI, Size; XXII, Weight; XXIII, Color; XXIV, Space; XXV, *Order?;* XXVI, *Time?;* XXVII, Number; XXVIII, Tune; XXIX, Language; XXX, Comparison; XXXI, Causality; XXXII, Wit; XXXIII, Imitation of the basis of the mind is hardly compatible. More congenial to that view would be a picture of the brain as a rather amorphous structure serving simply as the locus for the effects of the soul on the body.

Results accumulated in the late nineteenth and early twentieth centuries supported the basic correctness of the localist view. As researchers probed the question in more detail, it became evident that very specific injuries can result in correspondingly specific functional impairments, not just general degradation. And in this connection, the study of language played a central role.

Nowadays we do not seriously doubt that brain functions are substantially localized. We have seen that evidence from a number of sources makes it clear that in normal, right-handed adult males the bulk of linguistic function is in the left hemisphere. It would be interesting, of course, to be able to locate language more precisely. We might try, say, to identify the particular collection of neurons in which our knowledge of the word "grandmother" is represented.

The search for that kind of link between brain tissue and language is surely misguided. Even the localization that we do find is highly variable across otherwise comparable individuals. In a study of over a hundred preoperative epilepsy patients, researchers found that stimulation in precise locations could interrupt the ability to name objects. That may sound promising, but apart from the fact that by and large the relevant sites were in the perisylvian region, they varied tremendously from one individual to the next.

The connection between language and the body is evidently abstract, and the language organ is a functional notion, not an anatomically localizable organ like the kidney. All the same, the details of language function are tied, even if indirectly, to the details of our organic constitution. And that means that the development of language in the child should be seen as a matter of maturation and development—"growing" a language, not "learning" in the sense in which we learn the state capitals. In this respect it is entirely parallel to similar developments in other animals, such as song in birds.

The Ontogeny of Human Speech

Let us grant that the human brain has specialized ways of dealing with language, consequences of the details of its neurophysiological structure as it develops in childhood. The course of this development is governed in part by the capacities that the organism's genetically determined organization provides. Specific choices within this range are determined by an interaction with experience and the environment.

As a schema, this picture seems to fit well with what we know of the development of speech in children, and also of song in baby birds, together with much else that goes by the name of "learning." Let us now turn to a somewhat more circumstantial account of the course of early language acquisition, confining ourselves to speech production and perception.

Looking first at speech production, we find that children follow a regular progression during the first year or so of life, independent of the culture in which they are raised. Until the age of about five months, as they make their first attempts to control the apparatus of speech, they produce vowel-like sounds generally referred to as *cooing*.

Around seven months, they begin the stage known as *babbling*. Early on, it consists of repetitive productions that rhythmically alternate consonants and vowels ("babababa," "mamamama"). By around eight to ten months, the vowels produced in babbling begin to approach those specific to the language spoken by the parents and others around the children. They also begin to alternate different syllables rather than repeating the same one. The intonation contours of babbling start to resemble those of their (soon-to-be) native language, and by the age of about ten months it is possible to differentiate children on the basis of the linguistic environment in which they have developed.

This development is a crucial part of the infant's attaining control of language, as we can witness from the behavior of congenitally deaf babies. Until the onset of babbling, the vocalizations of deaf infants are entirely comparable to those of hearing babies, but deaf babies do not continue to babble. Around the age of seven months, their vocalizations diminish and do not reemerge until later. At that point the sounds are dominated by syllables with labial consonants that the baby can *see* how to pronounce.

The absence of vocal babbling in deaf infants does not imply that this stage of language development is absent. For deaf children raised among signing adults, there is indeed a linguistic environment: it is simply in a different modality from that of hearing children. Laura Petitto and her colleagues have shown that deaf babies do engage in a manual "babbling" that evokes the constituent elements of signed language. Qualitatively different from the manual activity of hearing children, it serves the same function of attunement to a linguistic world that oral babbling does for the hearing child. Indeed, this manual babbling proceeds through several stages that are strikingly parallel to those observed in the vocal babbling of hearing babies.

By around ten to fifteen months, hearing children have arrived at a selection of vowel and consonant types appropriate to their native language. Babbling may persist, in the production of nonsense repetitions (generally with appropriate sentential intonation), but stable words begin to appear by the end of the first year, and infants now come to use a consistent phonetic form to refer to an object. Around twenty to twenty-four months, when most (though by no means all) children have a vocabulary of roughly 250 to 300 words, they begin to combine words in meaningful ways and produce their first sentences.

A child's early productions are of course available for observation, and many studies have documented the path of development sketched above. It is naturally more difficult to study the sequence of events in the development of perception, since we cannot directly observe what is going on in the mind of the prelinguistic child. Recent years, however, have seen the emergence and refinement of ingenious experimental techniques for drawing conclusions about a child's perceptual abilities, and we do know quite a bit about the sequence in which linguistic perceptual ability arises.

We can show that even before birth, the fetus responds to auditory input, recognizing changes in sounds and reacting preferentially to the sound of the mother's voice. At birth (or as soon thereafter as it is practical to conduct experiments), infants can discriminate sounds along all of the dimensions used by the world's languages. They also can detect changes in intonation patterns, while recognizing the sameness of a speech sound despite variations in intonation.

At or shortly after birth, infants show a preference for the prevailing intonation patterns of their "mother tongue." While initially startling, this fact does not necessarily tell us much about a language-specific facility. Tuning in to this kind of property of speech can rely on general characteristics of the auditory system that are common to nearly all primates. Not only newborn human infants but also cotton-top tamarin monkeys can discriminate sentences from Dutch and from Japanese (though not, in either case, if the sentences are played backward).

Up to the age of two months, infants show no right-ear advantage for speech (they do show a left-ear advantage for musical contrasts). By three or four months, though, a right-ear advantage for speech emerges. Its significance lies in the fact that the primary language areas in the brains of most humans are in the left hemisphere, to which the right ear provides the most direct access.

By the age of five months, infants can make some connections between visual and auditory information. Around six months, they show a preference for speech containing the vowel sounds of the language spoken around them. At about the same time, they are able to detect the prosodic cues for the boundaries of clauses in different languages, although it will be some time before they actually produce utterances that could be said to be structured into "clauses."

Around eight to ten months, sensitivity to prosodic organization increases, and babies can be shown to be sensitive to phrase boundaries within clauses. This ability is obviously critical if they are to be in a position to impose a syntactic organization on their linguistic input (which we must assume if their knowledge is to include the kind of syntactic regularities discussed in Chapter 8). Although they do not yet produce stable words, infants can be shown to prefer word forms that respect the stress patterns of their native language, and the constraints on sequences of sounds that are characteristic of it.

Already at the age of ten months, infants' ability to discriminate sound contrasts that are not used in the language spoken around them begins to decline. This degradation of the phonetic perceptual virtuosity they are born with will continue as they acquire their native language. Adults have considerable difficulty hearing sound contrasts not present in their native language. Japanese speakers cannot easily distinguish [r] from [l], English speakers have trouble hearing French or Spanish [p] as opposed to [b], contrasts that all could discriminate as babies. When the same physical acoustic dimensions are presented in a nonspeech signal, adults perceive them with roughly the same accuracy regardless of native language. It is specifically *linguistic* perception that becomes preferentially tuned to the dimensions of contrast utilized in a particular language.

These observations suggest the following course of development. At birth, the child's perceptual system is capable of a range of auditory discriminations. Both speech and general auditory processing have roughly the same capacities. During the first months of life, the quality of the speech around the child results in a "tuning" of the speech perception system that begins to focus on the kinds of sound found in the surrounding language. At about six months, the child begins to experiment with his own vocal apparatus, discovering the connections between articulation and sound. As the perceptual system becomes tuned to a specific language, and integrated with the motor system, it comes to disregard distinctions that do not occur in that language. This process has no effect on nonspeech perception, however.

This progressive refinement seems to proceed as a product of the availability of evidence. The child in the womb has access to some limited information about speech, including the rhythm and intonation pattern of the mother's speech but little or nothing else. We are able to demonstrate some specialization for those patterns right after birth. (Corresponding results exist for birds: prior to hatching, baby birds hear the vocalizations of their parents and prove to have a preference at hatching for similar patterns.)

It has long been believed that this sort of development takes place during a specific stage in life: the critical, or sensitive, period. For some aspects of language learning, we have actual evidence that if the correct kind of stimulus is not present during the appropriate age period (roughly, up to puberty), learning does not take place. The classic example is the complex and troubled case of Genie, discussed in sympathetic detail by Susan Curtiss. Although other relevant examples appear in the literature (particularly cases of deaf children who were not exposed to signing in early life), little information exists on the development of perception. The few known studies of children deprived of linguistic input during their early years apparently did not include the kind of perceptual tests that would inform us about their speech perception systems.

It is possible that in "late" bilinguals (people who learn a second language well, but after childhood) as opposed to "early" bilinguals, processing differences occur. Imaging studies seem to show that the latter use overlapping brain regions to process the two languages, while the former use separate, distinct regions. Perhaps whatever we learn after the period of normal first-language acquisition, we learn in a different way.

Why should this be true? One possibility is that it is just an aspect of development: certain mechanisms cut in at a particular point and cut out at a particular point. A slightly less metaphysical point of view is the following. In the early years of life, the brain develops vast numbers of new synapses, peaking at about two years of age. Over the next several years, neurons die and synapses wither, resulting in a specifically limited organization. Neuron death levels out at around age seven, whereas overall synaptic organization is determined by about the age of puberty. If we assume that the role of experience is to shape and mold this specific organization, then whatever is in place by puberty is what we have for life.

Certainly, little or no overlap occurs in the details of the development of speech in children and of song in birds. Equally obvious, however, is the remarkable similarity of these two processes at only a modest level of abstraction. Both kinds of infants begin by absorbing general properties of the models provided by experience during a specific, limited sensitive period. Both pass through an early stage in which they establish systematicities in the relation between what they can do with their vocal apparatus and the sound that results. This activity (subsong in birds, babbling in human infants) undoubtedly prepares the underpinnings for the specialized, motor-based perceptual systems they both employ as adults in dealing with the characteristic vocalizations of their species. And both then proceed to try out the range of possibilities consistent with the Universal Grammar of their species, settling eventually on an appropriate approximation to the system found in the world of their conspecifics.

We should have little hesitation in seeing both processes as essentially similar, as the working out of a species' developmental program of biologically guided maturation. In other words, nestlings and babies both grow up in a specific way, determined in its essence by the fact that they are birds and humans, respectively.

7

What Primates Have
to Say for Themselves

And many of the tales that Chee-Chee told were very interesting. Because although the monkeys had no history-books of their own before Doctor Dolittle came to write them for them, they remember everything that happens by telling stories to their children. And Chee-Chee spoke of many things his grandmother had told him—tales of long, long, long ago, before Noah and the Flood—of the days when men dressed in bear-skins and lived in holes in the rock and ate their mutton raw, because they did not know what cooking was—having never seen a fire. And he told them of the Great Mammoths and Lizards, as long as a train, that wandered over the mountains in those times, nibbling from the tree-tops. And often they got so interested listening, that when he had finished they found their fire had gone right out; and they had to scurry round to get more sticks and build a new one.

— The Story of Dr. Dolittle

The birds, the bees, the frogs, and the other animals we have looked at all can teach us. If we hope to find communicative behavior anything like our own, though, surely we ought to look at animals closer to home: the (non-human) primates.

If we are interested only in finding some anticipation of human language, the resulting survey is largely discouraging. Primate communication is qualitatively similar to that in other animals, and while intriguing and novel features are present, they do not in the end point to systems that are interestingly closer to English than, say, bird calls. We can certainly learn a great deal, though, by asking how our primate cousins communicate in their natural circumstances.

Compared to bird calls, monkey (and ape) vocalizations, as systems, are rather similar. They comprise a limited range of signals, each of which seems to express some aspect of the animal's internal state. The set of signals is innately determined, though in some special circumstances, limited kinds of learning affect the ways in which they are used and interpreted. In a few cases, we have evidence that individually simple signals can combine to yield a new signal whose import is not identical with any of its parts. The possibilities for such combinations are extremely limited and do not yield any sort of productive system that serves as the basis of free innovation.

In some instances, a connection exists between the choice of a vocalization and some aspect of the external world (the presence of one specific sort of predator as opposed to another); some researchers would say that the calls in question *refer* to eagles and leopards, much the way words of a human language refer to things. If this is indeed the right way to characterize these cases, the role of reference (or any other relevant property of human languages) remains both limited and rare. Furthermore, it may turn out that thinking of these signals primarily as references to things in the world will actually obscure our understanding of some aspects of their structure.

Natural communication among primates can be based on vocalizations or on gestures. A difference in biological classification coincides to a surprising extent with this distinction: monkeys tend to have richer systems of vocal signals but use fewer significant gestures, while the opposite is true, more or less, of the apes. These are broad generalizations, of course, but they explain why our discussion of vocalizations focuses on monkey species such as vervets and rhesus macaques, and not on the chimpanzees, gorillas, and bonobos that are of interest when we turn to gestural signals.

We begin with a look at the alarm calls given by vervet monkeys (*Cercopithecus aethiops*) upon sighting a potential predator. Distinct calls are given for distinct types of predator, and these elicit distinct types of appropriate responses in the caller's fellows. This somewhat remarkable fact has produced a large literature on the question of whether the calls are genuinely referential — and it has also stimulated related studies on the alarm behavior of a variety of other animals. The example of the vervet throws interesting light on the superficially similar, but structurally rather different, behavior of other animals such as chickens, ground squirrels, and marmots.

Researchers have also studied alarm calls of several species of lemur, a primate somewhat more removed from us although within our order. Lemurs produce a number of other communicative vocalizations, and these are worth investigating as well. Perhaps even more interesting is the lemur's use of a completely different channel, that of olfaction, in structured ways to get across messages that are central to the animal's way of life.

Of course, it is not only prosimians such as the ring-tailed lemur that use sound for purposes other than sounding an alarm. Further study of the vervet monkey reveals a range of other calls beyond those that first made the animal famous in the ethological literature. Close observation has taught us a tremendous amount about the cognitive world of the vervet on the basis of these other vocalizations. In fact, the clearest evidence for the referential nature of vervet vocalizations is perhaps to be found in the properties of calls that are related to the social organization of vervet groups.

What of our closest relatives, the higher apes, such as the chimpanzee and the bonobo, the gorilla and the orangutan? These animals will be of particular interest to us in Chapter 10, where we ask about the extent to which they can be taught to use a human language under special conditions. What do they do on their own by way of communicating? It would be fascinating to discover the precursors of human language in their behavior, a result that some argue would be absolutely essential if we are to confirm the continuity of evolutionary sequence.

That does not seem to be the case, although laboratory studies *do* reveal cognitive capacities in apes that seem to be lacking in monkeys. It is an extremely interesting result, not only in its own right, but also for what

it shows us about how to interpret the skills that can be elicited from these animals in the laboratory.

But we are getting ahead of our story. Let us begin by looking at the vervet monkey's most famous accomplishment, the ability to let others know not only that danger impends, but also something of the nature of that danger.

Alarm Calls

It was Tom Struhsaker who, in 1967, drew the attention of students of animal behavior to the vocalizations of vervets. These little "Old World" monkeys are found widely in savannah, forest, and semidesert parts of sub-Saharan Africa, where they form one of the most common primate species on the continent. They are rather small in size (adult males weight about ten pounds, females perhaps eight) and subject to a variety of predators. Fortunately for their survival, they have a corresponding variety of ways to protect themselves.

The animals that prey on vervets fall into three general classes. First are the large cats, primarily leopards, that chase the monkeys on the ground. The dexterity of the vervet provides a great advantage in climbing, and an effective escape strategy when pursued by a leopard is to climb a tree.

A second class of predator would be more than happy to see the vervet employ this strategy. The martial eagle (along with two or three other large birds of prey) is likely to swoop out of the skies to carry off a vervet for lunch. If the target were to climb a tree, that would make the eagle's task much easier. When an eagle is sighted, therefore, the best thing for a vervet in a tree to do is get down and hide in the bushes.

Finally, a variety of snakes, including pythons, mambas, and cobras, are dangerous to vervets, but only if they catch them unawares. In the presence of these it behooves vervets to pay constant attention to their location, which makes evasion relatively easy.

Struhsaker observed that when predators of one or another of these types are spotted by one of a group of vervets, different-sounding alarm calls are given depending on which type of danger is present. Leopards and other large cats elicit loud barks, or "leopard alarms"; upon seeing an eagle, the monkey gives a characteristic kind of cough or "eagle alarm"; while a third, acoustically distinct sound called a "chutter" or "snake alarm" is produced when a python or other dangerous snake is seen.

Even more fascinating than this differentiated calling is the response of the other vervets. When one of them gives a leopard alarm, the others bark loudly and run for a tree, if on the ground; or climb higher, if already in a tree — regardless of whether they themselves can see the leopard. Similarly, when one member of the group gives an eagle alarm, the others climb down from trees they are in and rush to the bushes. A snake alarm results in all of the other monkeys standing up on their back legs and looking around in search of the snake. They may even approach and mob it, giving further calls (from a safe distance) to ensure that everyone knows where the danger is.

In each case, the response to the alarm call is appropriate to the kind of danger the predator presents. The easy and obvious interpretation, for the human observer, is that the calling monkey is shouting "Leopard!" (or "Eagle!" or "Snake!") and the others are simply acting sensibly on the basis of this information. But is this anthropocentric story the right one? Is the monkey really "referring" to a specific animal? Does the call simply reflect a kind of fright? Do the other monkeys act on the information that a specific danger is at hand (to wit a leopard, an eagle, or a snake) or are they merely responding to their colleague's fright?

Philosophically, the difference between seeing the monkey's alarm call as having a specific external reference, on the one hand, or as simply reflecting the animal's internal state of fright, on the other, is a significant issue. To see that it is also behaviorally interesting, let us consider another, superficially similar case: that of ground-dwelling sciurids (chipmunks, ground squirrels, prairie dogs, marmots, and the like).

For a Belding's ground squirrel (*Spermophilus beldingi*), danger again comes in more than one form. An aerial predator such as a hawk may swoop down out of the sky; or a carnivore such as a badger, weasel, or coyote may stalk the animal on the ground. Like the vervets, apparently, a squirrel that detects one of these potential dangers gives one or the other of two acoustically distinct alarm calls, and the other animals respond in a more or less appropriate way (ducking immediately into a hole for the aerial predator and looking around watchfully for the other enemies).

When we look more closely, though, we see differences between the vervets and the squirrels. The vervet calls are all distinct from one another, but the squirrel alarms represent the two ends of an acoustic continuum, with intermediate forms also heard on occasion. More important, the squirrels often give what seem to be the "wrong" calls. When closely pursued by a carnivore, they may give an "aerial alert" call, and a hawk sighted at a great distance or standing on the ground may elicit a "ground alarm" call.

The consensus that has emerged about the interpretation of alarm calls by ground squirrels is that these vocalizations indicate not the specific nature of the potential predator, but rather the degree of urgency of response to the danger. Hawks can swoop out of the sky very quickly, and when one is around, it is essential to duck for cover immediately; the same is true when a fast-moving carnivore is in hot pursuit. When a coyote is simply stalking the neighborhood, the squirrels need to remain alert, but not necessarily to drop everything and run. The same is true if a hawk is barely in sight, not near enough to attack without warning.

Squirrels, it seems, have basically one thing to say ("Look out!"), and differences in the way they say it reflect their degree of concern. Vervets also reflect degree of urgency in their alarm calls, but they do so by calling longer and more loudly when the danger seems greater. A variety of ingenious experiments, primarily conducted by Dorothy Cheney and Robert Seyfarth, have established quite conclusively that vervets have (at least) three distinct things to say about clear and present danger, not just one.

Much of the work in support of this conclusion involves playing back recordings of naturally produced alarm calls to unsuspecting monkeys in the absence of any actual predator. When leopard, eagle, or snake alarms are played, regardless of the length or loudness of the call (and thus, the implied urgency of the danger), listening vervets climb trees, hide in bushes, or stand up and look around, as appropriate. Their activity suggests strongly that three distinct *categories* of danger are signified by different calls, not three degrees of imminence of impending disaster.

For a second line of argument in support of this conclusion, Cheney and Seyfarth ran a number of experiments in which they repeatedly played, for instance, the leopard alarm recorded from individual A to other members of the group in the absence of any leopards (or of A). After a certain amount of time, the other vervets stop running into the trees and begin to ignore these alarms. Vervet A has been shown to be unreliable as a source of information about leopards. Once this degree of distrust in A's leopard-calling behavior has been developed, other calls can be played, and several interesting results emerge. First, the monkeys respond to the leopard alarm calls of some other individual, B, quite normally. And second, even A's calls are responded to when they involve eagles or snakes. That is, the fact that A is demonstrably to be ignored on the subject of leopards does not in itself compromise his evidence about other predators, or other animals' calls indicating a leopard.

The possibility that alarm calls from specific individuals can be identified as unreliable is not limited to vervets. Similar results have been obtained in work on Richardson's ground squirrels (*Spermophilus richardsonii*), a species whose alarm calls do not appear to depend on the identity of the predator as opposed to the urgency of the threat posed. C. N. Slobodchick-off has claimed that the alarm calls of another sciurid, Gunnison's prairie dog (*Cynomys gunnisoni*), not only vary depending on the kind of predator (hawk, coyote, human), but also that "within a predator category, the prairie dogs can incorporate information about the physical features of a predator such as color, size, and shape." If more research confirms these remarkable claims, it would mean that these prairie dogs have the most sophisticated system of vocal expression in the animal world, apart from human language.

Further research in vervet communities turns up other alarm calls as well. There may be one for mammals that are less obviously dangerous than leopards, but worth keeping an eye on; one for baboons, in places where those apes prey on vervets; or one for "unfamiliar humans," especially the Maasai tribesmen of the area. In some (savannah) areas of Cameroon where vervets are hunted by feral dogs, the dogs are treated as "leopards." Elsewhere, however (in forests), the dogs that might be encountered are working for human hunters; shouting a lot and climbing into a tree would not help in evading the real danger. What this situation requires is sneaking off into the bush. And indeed, these monkeys have developed another call, a "dog alarm." It is short and quiet, and on hearing it the other vervets sneak off into the bush where hunters cannot follow.

Sharing the News

A common feature of alarm calling is that when multiple species inhabit the same area, they are often able to make use of one another's alarm calls. This is nicely illustrated by the vervets, in areas where they share territory with animals that have their own systems. For instance, superb starlings also have aerial and ground predator alarms, and vervets in these areas respond to starling alarm calls in ways appropriate to the birds' own categories. A starling's aerial alarms may make the vervets climb down and head for the bushes, while terrestrial alarms may send them into the trees.

The predators that pose a danger to starlings are not all sources of concern to vervets. As a result, it is reasonable that the extent to which vervets pay attention is related to the closeness of the starlings' categories to their own. Vervets seem sensitive to the fact that starlings' calls do not have quite the same "meaning" as their own. For instance, the starlings' aerial predator call is given for a broader range of raptors and other birds. When they hear it, vervets tend to look up, but not to automatically run for the bushes. The starlings' terrestrial predator call is much more comprehensive than that of the vervets, and is given not only for leopards and snakes but also for other dangers that vervets do not worry about (like other vervets!). While they pay attention to this call when they hear a starling produce it, they do not treat it in exactly the same way they would treat the leopard call of a conspecific.

These facts confirm the impression that vervets have categories for different sorts of predators, and their calls reflect these. They also show that some form of learning is relevant in relation to these systems. That must be the case if the calls of other species (which bear no particular acoustic resemblance to the vervets' own calls) can be associated with the same categories.

This conclusion is also supported by research on quite different animals. Among lemurs, species such as the ring-tailed lemur (*Lemur catta*) and Verreaux's sifaka (*Propithecus verreauxi verreauxi*) have alarm call systems in which the individual calls can be shown, by arguments like those above for the vervet, to designate particular types of danger. Typically, they distinguish aerial as opposed to terrestrial predator types. Other kinds of lemur, such as the ruffed lemur (*Varecia variegata variegata*), have alarm calls that differ, like those of the ground squirrel, only in terms of urgency.

Both ring-tailed lemurs and sifakas have distinct calls for two categories of danger, although the calls themselves are not acoustically similar in the two species. Nonetheless, ring-tailed lemurs from areas where the two are found together can be shown to respond appropriately to the calls of a sifaka when these are played back to them. On the other hand, ring-tailed lemurs from a group raised entirely in an animal park (in Japan) where there were no sifakas did not respond to these calls. They had never had the opportunity to learn that these other animals know how to say "eagle," too — although they say it differently.

Another kind of learning must be involved to the extent that particular groups of vervets come to produce new, contextually appropriate calls, like the baboon and dog alarms mentioned. In the case of feral hunting dogs, all that is involved is learning a new condition of use for the leopard call, but the other two alarms present a different problem. If alarm call systems are innate in origin — as I argue below — we need to account for the appearance of novel calls on the basis of environmental factors. One possibility is that the vocalizations are part of the animal's innate repertoire, and the innovation consists in conditioning their use under certain circumstances of danger. This suggestion is consistent with what is known about the acoustic structures of the calls concerned, but further speculation will have to await the collection of additional data.

Glimmerings of Syntax

Among the monkeys that make use of one another's alarm calls are two species that live together in the Tai forest in Côte d'Ivoire, Diana monkeys (*Cercopithecus diana*) and Campbell's monkeys (*C. campbelli*). Each has discrete alarm calls for leopards and for eagles. Male and female Diana monkeys have calls that are acoustically quite different, but each sex responds similarly both to its own calls and to those of the other sex. The difference between one call and the other depends universally on the nature of the predator, not on some other factor such as the urgency of the threat or the direction from which the predator is approaching. The alarm calls of the two species are distinct from one another, but all of the monkeys respond in essentially the same way to (a) an alarm call of their own species, (b) an alarm call of the other species, or (c) the vocalizations of an actual predator. They reply with their own corresponding calls and take evasive action appropriate to the predator in question. If this were all there were to the situation, we would simply have an unambiguous example of the sharing of alarm calls across species.

There is a further twist, however. In situations where the danger is not very great, male Campbell's monkeys emit a pair of low, resounding "boom" calls before their alarm calls. The boom calls may be given when a large branch breaks or a tree falls, or when the monkey hears a distant alarm call, or when the predator is sighted at a distance. In each instance the level of danger is low or completely inferential, as opposed to the direct threats that elicit normal alarm calling. And as opposed to what happens when they hear a normal Campbell's monkey leopard call, when a Diana monkey hears a "boom"-introduced leopard call, they do not respond to these more complex vocalizations with calls of their own or evasive action. Apparently they know that this sequence indicates the mere possibility of danger, not its imminence.

A series of playback experiments made it possible to explore this alternative a bit further. The normal eagle and leopard alarms of Campbell's monkeys were played through speakers, and the Diana monkeys responded as expected to each. When a boom sequence preceded the exact same call by about twenty-five seconds, however, the Diana monkeys simply went about their business. It seems they interpret the boom as something like a modifying "maybe," so "boom—leopard" means "maybe a leopard." Furthermore, the Diana monkeys recognize this message specifically in the context of Campbell's monkey calls. Playbacks of Diana monkey alarm calls always elicited responding calls and evasive action, even when these were preceded by the same boom that mitigated the force of the Campbell's monkey calls.

The situation is most unusual, not only for alarm calling, but more broadly for signals in animal vocal or gestural communication. What is striking is that the Campbell's monkey calls, at least as interpreted by the Diana monkeys, involve the combination of two distinct signals (the boom and the normal eagle or leopard call) to create a new unit that means something other than what either signal means by itself.

The amount of "syntax" we can see in this example is extremely limited. The combinations consist of exactly two elements, one of which is fixed (the boom) and the other of which is taken from a restricted set (other alarm calls, for eagles or leopards). As I argue in Chapter 8, such a principle of combination displays none of the major syntactic properties of human natural language; in particular, it is neither hierarchically organized nor recursive in nature. Nonetheless, it is the closest we have come to finding a system involving anything like syntax in animal communication.

Sound Production and Perception

If we want to understand the vocal behavior of other primates, a reasonable place to start is with the systems they have available for sound production and perception. Fortunately, most of what we already know about the corresponding systems in humans will suffice, since primates are all rather similar in these regards. The main differences follow from the fact that we have some specializations for speech that other primates lack.

Human speech production, as we saw in Chapter 5, can be analyzed as involving (a) a source of acoustic energy, and (b) a filter that variably affects this energy so as to determine just what sound comes out. We generate acoustic energy in two ways: either as a series of rapid puffs in the airstream coming from our lungs (resulting from laryngeal vibration) or else as noise produced by turbulence at a point of constriction in the vocal tract (in the case of fricative sounds such as *s*, for example). We have seen that birds have a different system, with noise produced either by vibration or by a kind of whistling at the two outlets in the syrinx, but other primates are like humans in relying on laryngeal vibration and constriction turbulence as sound sources.

There are many kinds of nonhuman primates, of course, from bush babies through gorillas, and there are differences in the anatomy and physiology of the vocal organs. As far as the basic source of acoustic energy is concerned, however, the differences are quite minor. Prosimians, monkeys, and apes all have larynges quite like ours, and they make their vocal folds vibrate by essentially the same mechanism we use.

One difference is worthy of note, however. In humans, the vocal folds are stiff at the edges, and the tension there is rather precisely controllable by the intrinsic muscles of the larynx. Furthermore, although this edge is formed by a (membrane-covered) ligament, it is part of the same mechanical system as the adjacent muscle tissue. In monkeys and apes, the edges of the vocal folds constitute a somewhat more loosely coupled vocal "lip."

The first consequence of this dissimilarity is that the stiffness of the vibrating part of the folds is not as precisely controllable as in our own larynx, so when the vocal folds vibrate, the vibratory cycle is far less regular than in humans. As a result, the signal is less like a pure tone with regular harmonics, constant over time. The actual balance among harmonics (as well as frequency) fluctuates somewhat, of necessity. This variability does not seem to matter much to the monkeys; they do not make distinctions that depend on precise control of this structure.

The second consequence is that the loosely coupled vocal lip can, under some circumstances, vibrate by itself without needing to engage the rest of the folds. This mode of vibration can thus be at a much higher rate than if everything moved together. Some species can achieve fundamental frequencies much higher than those of the most talented soprano (up to several thousand hertz).

With respect to the filtering function of the supralaryngeal vocal tract, the differences are more significant, despite the superficial similarities between our anatomy and that of other animals. For one thing, nonhuman primates cannot exercise fine control over the configuration of their vocal tracts in the way we do. Therefore, they do not produce sounds that systematically require turbulent noise generation as an acoustic source. Monkeys do not hiss at one another, for instance, because they cannot manage to form the constrictions that are involved in the production of human consonant sounds.

In all of us (humans and other primates), the sound that comes out is not determined by the character of the acoustic energy source alone. The sound also reflects the filtering by the rest of the vocal tract. Speech, as we have seen, is primarily controlled by our ability to manipulate the organs of speech so as to change the resonances of the cavities between the larynx and the lips. We alter these over a wide range by quickly and precisely moving our articulators (the parts of the tongue, the velum, the lips themselves). The result is a diverse array of distinct sounds integrated into larger units (syllables, in particular) that emerge at an impressive rate.

In several ways, other primates are limited in their ability to do as well. First of all, their nasopharynx is normally coupled into the vocal tract. It is therefore difficult (though not impossible) for them to make sounds (based on a laryngeal sound source) that are not nasalized to some degree. Whistling sounds are based on turbulence generated within the oral cavity and thus escape this limitation, as do certain explicitly nonnasal vocalizations in some species, but most primate vocalizations implicate the resonant properties of the nasal cavities. Since these have a major effect in obscuring other aspects of the spectrum of sounds, the relative lack of fully nonnasal sounds would handicap an animal considerably in terms of phonetic range.

More important, perhaps, the human vocal tract differs from that of other primates in taking more or less a right-angle turn, thus dividing it-self into the oral cavity proper and the pharynx. Within this space, limited movements of the tongue can have considerable effects on the resonant properties of the whole system. In other animals, the vocal tract is configured in what is effectively a straight line, such that even if they could move their tongues as quickly and flexibly as we (which they cannot), the effects would not be as significant.

When we look at vocal communication in other primates, we find that differences among their calls are typically based on different patterns of sound-source production. These involve varying the fundamental frequency and/or the on-off temporal pattern of vocalization. Humans, in contrast, tend to phonate with a more constant source, allowing the varying resonances of the upper vocal tract to do most of the shaping of the sound that emerges. For example, in the sentence (much beloved by phoneticians) *We were away a year ago,* the larynx vibrates continuously and *all* of the differences are due to articulations of the tongue and lips. Of course, we vary fundamental frequency too (in the service of intonation), but this modification is relatively less important than the articulatory activity of the vocal organs.

The fact that nonhuman primates (like other animals) do not manipulate the formant structure of their vocal production in the dynamic way humans do certainly does not mean that their calls do not have formant structure, or that they are not sensitive to it. Formants are a mechanical and necessary property of any resonator. One needs only to have a vocal tract, not to be human, to produce sounds with this sort of structure. In nearly all animals, however, formant structure primarily acts as a cue to basic, invariant physical properties such as overall body size. Animals tend to manipulate the formant structure of the sounds they produce not for purposes of more precise communication, but to convey an impression of greater size and viability.

Nonhuman primates do alter their vocal tract resonances to some degree. The snake and eagle alarm calls of vervet monkeys differ not only in the quality of the noise source, but also in the spectrum of the resulting sounds as shaped by the vocal tract. Although one call differs from another in formant structure, there is no evidence of variation in vocal tract shape during the production of the call that is comparable to the constantly shifting nature of human speech articulation. Dynamic changes in vocal tract shape during the production of a single vocalization have been found in the gelada monkey (*Theropithecus gelada*), but these are isolated effects, as far as is known.

The nonhuman primate vocal tract at rest is not as long, comparatively, as that of humans, a fact that also limits the range of articulatory possibilities. The human larynx is permanently located quite low in the throat. It was once thought that this position was unique to humans, but we now know that many animals, including a number of primate species, can lower the larynx so as to produce a sound whose spectrum is consistent with production by a larger animal. *Homo sapiens* remains the only primate species in which this lowering has become permanent, presumably in large part as a specialization for the flexible articulation demanded by human speech.

The overall result of these differences is that nonhumans (primates and other vertebrates) cannot easily rely on vocal articulation to produce a significant variety of sound types. As a consequence, they have not developed the fine motor skills involved in such articulation.

The main articulator that nonhuman primates control is the lips. Protruding the lips causes the vocal tract to be longer, and the resonances lower. Spreading the lips and pulling them back (as in a grimace) effectively shortens the vocal tract and raises its resonant frequencies. Closing (or at least narrowing) the lips causes all the resonances to drop. These manipulations provide a limited amount of control over the filtering function of the tract. To change those properties significantly, an articulator such as the human tongue is required, but we have little evidence for the tongue as an active articulator in prosimians, monkeys, or apes.

Let us now turn from sound production in nonhumans to perception. All primates have essentially similar auditory systems, at least to a first approximation. As in humans (and birds), we can regard the ear as a kind of spectrograph that provides a spectral analysis of incoming acoustic information to the brain via the cochlea and the fibers of the auditory nerve.

Our questions about perception do not concern the structure of the peripheral auditory system itself, but rather what the animal's brain does with the information that system provides. In humans, we have reason to believe that the processing mechanism is specific to language, and located (usually) primarily in the left cerebral hemisphere.

Major differences between human and nonhuman vocal communication derive from the parts of the brain that are responsible for its control. In the squirrel monkey (*Saimiri sciureus*), at least, vocal production is not apparently under the primary control of cerebral cortical areas, but rather of subcortical areas more closely related to the limbic system. If we lesion cortical regions that are homologous to Broca's and Wernicke's areas in humans, we do not encounter gross abnormalities in call production. In contrast, the best-known forms of aphasia in human speech are the result of damage precisely to these areas.

Still, there does seem to be a specialized mechanism for *processing* conspecific calls that is at least partially lateralized. Lesioning the left auditory cortex impairs this perceptual mechanism (for a time; recovery is fairly rapid). Lesioning the right auditory cortex has no effect, while lesioning both (naturally enough) produces a deficit from which recovery does not take place.

Consistent with this picture of a special perceptual system based in the left hemisphere, studies with rhesus monkeys showed that the monkey preferentially turns the right ear (which is connected to the left hemisphere) to hear conspecific calls, but the left ear (linked to the right hemisphere) to listen to other auditory inputs. Thus, conspecific call perception may be localized in the left hemisphere as is much human speech processing. (The processing carried out by the monkey's system seems to be quite different.)

In connection with the motor theory of speech production raised in Chapter 5, monkeys provide some interesting evidence. Areas in the monkey's brain are active when the monkey performs certain actions with its hands, such as grasping an object—and *also* when the animal observes an experimenter performing the same actions. This activation is not seen, however, when the object is contacted by a nonhand structure. The neurons that display this Janus-like connection between production and perception are known as mirror neurons, and their analysis is one of the hot topics in neuroscience today.

Interestingly, the area where these mirror neurons are found is exactly the region (from the point of view of comparative neuroanatomy and the architecture of the animal's nervous system) that includes the homologue of Broca's area! The question naturally arises, why there? Speculation centers on an original coordination of hand and mouth in feeding behavior. The concept of understanding what one perceives by matching it to something one can do is a more general mechanism, but one that is represented in the brains of other species in just the right place to make us consider its possible evolutionary connection to speech behavior in ourselves.

A final area of interest in the acoustic communication system of nonhuman primates is the possible role of learning. To what extent is relevant experience necessary for the animal to come to control the set of calls it eventually displays? As we have seen in the case of birds, quite similar animals can have systems that vary greatly in the relative importance of nature and nurture.

Although the issue is complex, it seem reasonable to conclude that no learning is necessary for a monkey or ape to produce its calls in roughly the appropriate circumstances. Monkeys have fully mature and accurate calls at a very early age (twelve weeks). Animals raised in isolation, or those deaf at birth (either naturally or as a result of experimental mutilation), nonetheless produce appropriate calls. It seems evident that auditory experience does not play a crucial role in the development of a production capacity.

Contradictory evidence hints at the possible influence of environmental input on call learning in macaques. Rhesus and Japanese macaques have somewhat different vocal repertoires, and one experiment suggested that rhesus macaques raised by Japanese macaque mothers sound more like Japanese macaques than they would otherwise, and vice versa. But another experiment showed no such effect. A group of rhesus babies all came out sounding like rhesus, not like their Japanese foster parents. In any case, these two species are too similar to provide a valid test, since their natural vocal repertoires already overlap.

Basically, there is no evidence that learning in a strict sense is involved in the production of monkey vocalizations. This thesis is consistent with the suggestion of considerably less cortical involvement in sound production in monkeys than in humans, as noted above. Instead, subcortical structures in the limbic system are more fully involved. That basic architectural point need not mean that experience has no role in development. Field studies indicate that the monkeys do learn a certain amount about when and under what circumstances to call. That is, while they do not need to learn the actual calls, experience seems to play a role in refining their notion of the circumstances under which the calls are, as it were, called for.

As an example, in Cheney and Seyfarth's work with vervets, it was obvious that infants had only a vague notion of what constitutes an appropriate stimulus for, say, an eagle call. Young monkeys gave this call for a variety of large birds, not only for the limited number of species that actually prey on vervets. On at least one occasion, an eagle call seems to have been triggered by a falling leaf! By the time vervets reach maturity, however, their sense of what constitutes an eagle, and when it might be appropriate to point one out, has been substantially refined.

This result shows that monkeys' ability to use calls in appropriate ways develops, within narrow limits, in interaction with experience. There is little flexibility in this kind of learning, however. When Japanese macaques are brought up with rhesus macaques or vice versa, the animals do not adapt to their environment by using their innately provided repertoire of calls in a way appropriate to the expectations of the other animals around them. Some learning does go on in these experiments as far as what the calls of others mean: the cross-fostering mothers learn quickly what their adoptees' calls mean, even if it is not what members of their own species usually mean by a similar-sounding call.

There is another sense in which learning takes place in the vocal communication systems of monkeys. To some extent, local dialects develop: monkeys who are around each other tend to sound alike. This kind of fine-tuning is not the same as if the calls were learned from experience in the first place, but it does illustrate some modification of the production system on the basis of experience.

Communication among Lemurs

Alarm calls are no doubt the best-known (and most studied) instances of vocal communication among nonhuman primates, but they do not by any means exhaust the possibilities. Various species also have other messages to communicate, and even other ways to communicate them. Let us now consider the vocal and nonvocal communicative behavior of lemurs, primarily the ring-tailed variety; later sections will deal with nonalarm calls in monkeys and in apes.

Vocalizations

We have noted that ring-tailed lemurs give distinctive alarm calls in the presence of various predators. The adults produce several different sorts of vocal signals when they perceive a threat. These include (a) *gulps*, produced on seeing or hearing carnivores, raptors, rapidly moving humans, and other potentially threatening objects; (b) *rasps*, given on seeing large airborne birds that might be hawks or the like; (c) *shrieks*, when seeing low-flying (attacking) large birds; and (d) *yaps* or *barks* in the presence of mammals that are potentially dangerous but to which the lemurs respond by mobbing the animal. These calls are sufficiently distinct from one another and used under different enough circumstances that we surmise that they refer to different sorts, not degrees, of threat. Ring-tailed lemur vocaliza-

tions are thus similar to those of vervet monkeys—as opposed to those of ruffed lemurs and squirrels, which indicate only the seriousness of a danger and not its nature.

Ring-tailed lemurs produce other vocalizations in dangerous situations. They make a variety of *clicks,* short simple broadband sounds, either singly or in series, with acoustic differences that depend on whether the animal's mouth is open or closed. These clicks seem to indicate a certain level of concern and wariness, and let other lemurs know the location of the individual making the sound. The difference between closed-mouth and open-mouthed clicks relates to the level of arousal.

Because these animals live in forest environments where they are often not visible to one another, such a signal can be vital in maintaining group cohesion. Clicks are thus used under circumstances where no predator is involved: when moving around in the trees, to keep together; when a mother wants to summon her children; or when groups encounter one another.

Clicks are not the only sounds these lemurs make that have a function within their groups, which consist of perhaps thirty animals. They also produce antiphonal choruses of *moans* or *mews,* a kind of vocal exchange within a calm and relatively stationary group; as well as *wails* and *howls,* two very different calls that sometimes function together to help lost members find their companions. Solitary males may howl to attract the attention of receptive females, who may respond by wailing. Other vocalizations are associated with maintaining contact within the group, with the contentment associated with grooming and being groomed, and with other ordinary daily activities.

Still another collection of calls (*yip, cackle, squeal, twitter, plosive bark,* and *chutter*) is associated with the maintenance of dominance relations. In ring-tailed lemur society, all (adult) females dominate all males, and each sex has a fairly clear hierarchy. Individuals use vocal signals to express dominance or acknowledge submission, to challenge or threaten others, and so forth.

Without going into further detail, we can see that the ring-tailed lemur has a repertoire of more than twenty calls. Each has rather specific acoustic properties and conditions of use. In at least some cases (alarm calling, in particular), it is possible to see the calls not simply as expressing the lemur's internal state but as containing a degree of reference to things in the world.

So far we have been talking about the vocal signals used by adult lemurs. Infants have a different, more restricted inventory of calls: around a half dozen signals are used to attract and maintain contact with others in the group (especially the mother), and to indicate distress and discomfort of various degrees. The calls characteristic of infants appear quite early in life, with adult call types emerging essentially at the point where they might have a function for the individual. Without further evidence it is difficult to assess the role of experience and learning in the development of the lemur's call system.

We *can* say that, among primates, ring-tailed lemurs display a comparatively large range of vocal signals. If these signals were the animals' only ways of communicating with others, we would still say that they have a rich communicative life.

Olfactory Communication

Another, completely different system of communication is also relevant to the lives of these lemurs: communication through scent. The available documentation is sparse, and it is impossible to say a great deal with certainty about just what messages are conveyed in this manner, and how. The animals' behavior makes it clear, though, that this is an extremely important channel for at least some kinds of information.

Both male and female ring-tailed lemurs have glands in the anal and genital regions that produce powerfully scented secretions. These are used to produce scent markers on trees and other objects in the environment by rubbing against the object to be marked, and the markers are readily detectable by other members of the species. Individuals commonly place their marks not in random locations, but directly on top of a scent mark left by another animal. The earlier trace is not thereby obliterated, but it is certainly obscured.

Males also have scent glands under their arms. Besides leaving scent marks in much the same way as the anogenital glands, the underarm glands are used in a specific kind of combative display. Males competing for dominance rub their tails under their arms to impregnate them with scent, then wave them at one another as an act of intimidation.

The chemical content of these scent markers varies as a function of the individual, his or her current fitness and condition with respect to receptiveness to mating, and probably much more. Frustratingly, we know almost nothing about what information the lemurs might obtain from smelling these secretions. It must be substantial, though, for they devote a great deal of attention to leaving marks, sniffing at the marks of others, and countermarking them.

Lemurs are by no means unusual in the animal kingdom for their use of chemical signals, sent through the olfactory system, in the regulation of their social lives. Many other mammals, primates and nonprimates, leave messages for one another in this form either through the secretions of special glands or in other body products such as urine.

Since the olfactory cortex constitutes a more significant part of the brains of most animals than it does in humans, the detection of these scents undoubtedly plays a more prominent role in the animals' awareness not only of the world around them but also of their social environment than it does for humans. I have already mentioned (in Chapter 5) that mice have a specialized sensory organ and a distinct neural processing mechanism dedicated specifically to the ecologically important pheromonal signals of their species. In addition, research has shown that hamsters are crucially dependent on such signals for the regulation of their interactions with one another. They are, in consequence, extremely good at using chemical markers to extract information about the identity of other hamsters, including relatives they have never met.

Among insects as well, chemical markers play a critical role in regulating social structure, especially in identifying other conspecific insects as outsiders or members of their own colony or nest. A particularly intriguing story involves a species of tropical ant (*Ectatomma ruidum*) which appears to mimic the nestmate recognition pheromones of a different colony in order to enter its nest and steal incoming food. From an anthropomorphic point of view, we seem to have here a case of deceptive communication, like that of the piping plover mentioned earlier.

One theory of this behavior involves the ants' actively acquiring traces of the relevant recognition compound from the colony from which they wish to steal. While we need not assume that calculation and a theory of mind are involved on the part of the thief, the behavior would still be quite remarkable. Subsequent research suggests that what is actually going on is that a genetically determined "caste" within the colony is particularly suited to engage in this sort of larceny, because these ants have a significantly lower level of the substance that would identify them as members of their *own* colony, and so smell less like outsiders when they go past the guards of another colony. Whatever the explanation, this example reinforces the crucial role played by chemical signals in the social organization of other species.

While we have very little understanding of the structure of these olfactory messages, the factors which control that structure, or the perceptual dimensions involved in the detection and processing of their content, we can make a few general statements about the nature of such a system in comparison with other modes of communication. Harking back to the characteristics of communication systems cited in Chapter 2, we can see several ways (apart from their nonuse of the "vocal-auditory channel") in which chemical signals differ from those that are auditory or visual.

A chemical signal may have many components, but because all are present simultaneously, there is no internal sequential structure. In the case of the lemurs, where one animal deposits a scent on top of that left by another (and then another, and another . . .), the entire complex has a sequential nature. This structure is apparently something the animals can retrieve, but any specific chemical message is a unitary whole.

Animals may have little or no control over the content of these signals, only over whether or not to produce them (and sometimes not even that). When a scent mark is made, it is in effect broadcast to all who might be capable of detecting signals of this sort, not just to those who are in the immediate vicinity when it is made. Studies of lemurs suggest that most scent marks are investigated within minutes — or not at all. If the receiver did not happen to be present when the mark was made, its source may be hard to determine. And unlike sounds or gestures, chemical signals are not subject to rapid fading, but rather remain detectable for some time.

In these and other ways, the olfactory channel is very different, and somewhat less flexible, than other modalities for communication. Scent signals may be extremely expressive, however. An enormous amount of highly significant information about the identity, fitness, and other characteristics of individuals undoubtedly is transmitted in this way to those equipped to understand it. For regulating the social organization of groups of animals who may not be in constant touch by sight or by sound, scent signals can be quite efficient. In any case, they are obviously a biological specialization for communication—one that lemurs have, but that we humans, with our diminished olfactory cortex and sense of smell, generally lack. Every species has its own special systems, determined largely by its unique biological organization: language in our case, scent marks (along with vocal calls) for the ring-tailed lemur.

Probing the Minds of Monkeys

In vervets, lemurs, and other primates, alarm calls are far from the end of the story when it comes to communicative vocalization. These animals call when they find food (perhaps differently depending on whether the food is rare and desirable or just ordinary), in aggressive confrontations, during sexual activity, and at other times as well. The number of calls that can be differentiated depends on the species (and on the auditory acuity of the investigator), but does not seem to exceed one or two dozen. Some have been the subject of investigations that reveal a good deal about the mental life of the animals involved.

Like lemurs, vervets have a number of other vocal signals that play roles in the day-to-day social interactions of members of a group, with one another and with other groups. For instance, when one vervet group encounters another, a characteristic vocalization, a "wrrr," is given upon sighting. If the outsiders get close enough and actually make contact, the wrrr may be replaced by a chutter.

Within the group a lot of grunting goes on. This happens under four general circumstances: (a) when submissive meets dominant, (b) when dominant meets submissive, (c) when one monkey goes out into an open area, and (d) when a monkey encounters another monkey not of his own group.

Cheney and Seyfarth studied these intragroup grunts in considerable detail. They had the impression that they could more or less distinguish one kind from another, but failed in their initial attempts to find gross acoustic features distinctive of each. As a result, they considered the possibility that vervet grunting is always the same, the difference being purely contextual. A submissive animal knows his relation to a dominant, that is, as does the dominant, so there is no need for them to have different ways of expressing

that relationship. Perhaps a grunt is taken one way or another depending on the characteristics of the context.

But this is not the case; acoustic distinctions must exist among the grunts that are used under different circumstances. Cheney and Seyfarth recorded a number of grunts by different individuals in different situations, then played them back through hidden loudspeakers to others in various contexts. The four grunt types (based on the circumstances under which the original recordings were made) elicited qualitatively different responses. For instance, the "submissive-to-dominant" grunt caused the hearer to gaze toward the speaker, while the "other-group" grunt caused him to look out at the horizon. There must, then, be differences in the grunts, differences in the sounds that are interpreted by the monkeys, even if human listeners performing a detailed acoustic analysis find it hard to identify what those features might be.

In addition, these differences have an importance for vervets in their relations with one another. When some new males joined a group, for the first few days they and the members of their new group made other-group grunts to one another. After a while, though, the newcomers began to make submissive-to-dominant grunts to the group's dominant male, and the females of the group started making intragroup grunts to the newcomers as well. These were either submissive-to-dominant or dominant-to-submissive, depending on the individual's relative place in the group's pecking order—a matter that seems to have been resolved rather quickly. The different grunts do convey different sorts of information, then, and are not merely a single signal that can be interpreted in different ways depending on the context.

Proceeding on the assumption that the various calls really are different, we can ask whether they "mean" for the vervet what they seem to mean to us. One possibility is that these vocalizations are purely automatic, involuntary reflections of something—either of the vervet's internal state (perhaps alarm) or of some action the vervet is about to take.

In the case of the alarm calls discussed earlier, this last possibility is disconfirmed, because what the vervet actually does after giving the call varies. The monkey may do nothing at all, or may climb up a tree, or may climb down from a tree, without any necessary and inflexible relation between call and action.

If the call is not a reflection of impending action, what about the possibility that it reflects the animal's internal state fairly directly? This thesis would initially seem to be supported by the neurobiology: recall that vocalization in at least some monkeys appears to be initiated primarily in the subcortical limbic system, rather than in the cortex. But that notion will not suffice, for several reasons.

First of all, solitary vervets do not give alarm calls. While one might imagine that the presence of an audience affects the animal's internal state, this possibility somewhat reduces the attractiveness of our hypothesis. Studies of a variety of primates show that under some circumstances, the animals can apparently suppress their calling behavior—even while making the facial expression that would normally accompany it! Some element of voluntary control must be involved.

Second, dominant animals call more than submissive ones do. Why should this be? Perhaps because dominants are the ones doing most of the breeding. It is their genes that are more at risk to predation, so they care more about the group. In general, too, more alarm calling takes place in the presence of close kin or offspring than in the presence of peripherally related group members. This sensitivity to one's own relation to other nearby animals is seen in a great many other animals as well (witness the food and alarm calls of chickens).

In the particular case of vervets, all of these features demonstrate that alarm calling is sensitive (at least) to the audience, which means it cannot simply be a direct reflection of the monkey's internal state of alarm. Rather than being merely expressive, vervets perhaps produce alarm calls in order to influence the behavior of the other vervets, to get them to take appropriate evasive action with respect to the specific threat that is at hand.

Furthermore, the calls are sometimes given under circumstances that appear have the character of deception. On occasion when two groups were fighting and one group was clearly losing, some member of that group gave a leopard call. All of the monkeys ran for the trees and the fight stopped — but there was in fact no leopard, as least as far as the observers could determine. The calling monkey cannot have been frightened by a leopard if there was none. We might interpret his call as an attempt at deception, but remembering the cautions of earlier chapters, we must stop at the conclusion that the caller was attempting to get the other animals to disperse.

We can at least conclude that alarm calls are not just an automatic reflection of internal state, but are intentional. The monkey says "Leopard!" not because he is himself about to run into the tree, or because he has seen a leopard and is scared, but because he wants the others to take cover. Thus, the vervet is what Cheney and Seyfarth (following Daniel Dennett) call a *first-order intentional system.* The animal produces a signal by which he intends to influence the behavior of others.

Can we go beyond this, to establish that the calls constitute what Cheney and Seyfarth call a *second-order intentional system*? To show that, we would need evidence that a monkey intends to affect not just the behavior of another, but that animal's knowledge or state of mind. Here the evidence is disappointing. While the monkey's tendency to call depends on the audience, it does not seem to depend on the state of knowledge that members of that audience can be inferred to have. If the point of calling "Leopard!" were to make sure that everyone knew there was a leopard, one monkey would not need to call if other monkeys had already called, or could perfectly well see the leopard. That is not what happens. When one calls, the others also call, regardless. There is no evidence that they take one another's state of awareness of the danger into account in signaling.

Still, they clearly do identify the calls they hear with a specific calling individual. By repeatedly "crying wolf" with recordings played through hidden speakers, the experimenter can habituate members of a group to monkey X's leopard call, for instance; but monkey Y's leopard call will send them off to the trees. Reliability (or the lack of it) is associated with specific individuals, which means the animals must be able to make identifications on the basis of the voices they hear.

What should we conclude that the calls "mean"? We have already seen that they do not mean simply "I'm scared!" Should we say, then, that they actually refer to something in the world? On that interpretation, a leopard call would mean "There's a leopard!" or perhaps "There's a predator that attacks in a particular way!" The facts are certainly consistent with such an interpretation, but there is no evidence that forces that interpretation as opposed to the more conservative view that the call means "Go climb a tree!"

One consideration is more consistent with the "Go climb a tree!" interpretation than with "Leopard!" If the calls really refer to objects (predators) in the world, they are completely arbitrary. The leopard call bears no relation at all to real leopards. It does not resemble in any way the sound of a leopard or evoke one other than by convention. The same is true for the eagle and snake calls. Words in human languages are arbitrary in this way, of course, but the relationship in primates is something of a mystery.

No complete explanation of the form of the calls is available, but seeing them as attempts to influence the actions of the other monkeys (rather than referring to something in the world) helps somewhat with the problem. The acoustic structure of all three alarm calls is similar. All are broadband sounds, with very sudden onset and a rapid rise time. They are thus perfectly structured to evoke the "startle" reflex in hearers, to get their attention and provoke them to interrupt what they are doing. This general response is known in a variety of monkeys and apes. Correspondingly, the acoustic analysis of alarm calls shows that they usually have just the kind of structure that could elicit it.

Among the vervets, we find the exception that proves this rule. Recall that in some areas vervets have developed an alarm call for dogs that might well be in the company of human hunters. Unlike the leopard, eagle, and snake calls, this call does not have a sharp and sudden character, but is rather soft and acoustically diffuse. It is not nearly as easy for a listener to localize, and it is similar to calls in other species that evoke caution rather than sudden abrupt action.

If we think of the alarm calls in terms of their intended effects, their acoustic structure makes sense. If we insist on seeing them as referring to particular predators directly, however, that structure remains arbitrary and inexplicable. Further, we see that when calls are generalized, the generalizations make sense in these terms. In areas where the leopard call may also be a response to sighting a feral dog, that apparent ambiguity is because the appropriate evasive response in the two cases is the same. "Leopard" does not refer to a species in anything like the way we think of one, but rather in a more functional sense, in terms of a class of appropriate reactions.

The referential nature of vervet calls is suggested more strongly by another of Cheney and Seyfarth's experiments, one that did not involve alarm calls. We have seen that two distinct calls can have more or less the same reference. The wrrr and the chutter both refer to "another group." Cheney and Seyfarth played misleading wrrr calls long enough that some listener became habituated: he learned that the caller was giving a false alarm, because no other group was coming into view. The researchers then played a chutter and discovered that the listener ignored this call too—because both wrrr and chutter refer to the coming of a new group, and the calling monkey had already proved unreliable in that respect. Contrast this with the fact that an individual's unreliable eagle calls do not result in the other vervets' ignoring his later leopard call. These experiments are suggestive of a "referential" interpretation, since two distinct calls with essentially the same reference are seen to be equivalent, while two calls with different reference are not.

Much can still be learned about the social and cognitive life of monkeys from a close study of their communicative behavior, as Cheney and Seyfarth show. These animals have a clear sense of individual identities, social and family relationships, and much else, and act on the basis of a highly structured view of the world and their own places in it.

In our efforts to understand the nature of communication, the principal point seems to be that vervets have a system which they use with the apparent purpose of influencing one another's behavior. There is no evidence that they (or any other monkeys) have a theory of mind in the sense of an understanding that other monkeys have their own knowledge of the world, that this knowledge plays a role in determining their actions, and that one can influence another's behavior by affecting that knowledge. As a result, we can conclude only that vervets intend to modify one another's actions, not that they try to deceive or otherwise shape one another's beliefs.

Communication among the Higher Apes

Prosimians (such as lemurs) and monkeys communicate vocally using elements from a limited set of calls. These number up to perhaps a few dozen and are largely innate; their conditions of use can be modified by experience, however. The calls are only partially voluntary, although the brain regions under whose control they lie are not well understood in most species. At least some of these animals (lemurs in particular, and many monkeys as well) also transmit and receive information through scent marking and other olfactory channels.

The primates of greatest interest to us, *Homo sapiens*, make use of open-ended vocal (or manual) systems based on the recursive combination of discrete units into hierarchically structured forms with complex semantic interpretations. These systems are acquired from within an innately specified range on the basis of experience and (apart from the occasional "Ouch!") are used under the voluntary control of cortical mechanisms. The perfume industry notwithstanding, we make only minimal communicative use of the information that can be derived from chemicals in our environment.

Given these rather sharp differences, we might expect that our nearer relatives among the primates, the great apes, represent some sort of intermediate position. We might look to them for communicative means that, while perhaps not really *language* in our sense, would still be more like language than the calls of lemurs and monkeys. In this we will be disappointed. From all existing studies, communication among the great apes seems qualitatively little different from that among the other nonhuman primates.

Vocalizations

Although there have been a few studies of communicative behavior among gorillas, orangutans, and bonobos, the great bulk of the literature concerns chimpanzees—specifically, Jane Goodall's extensive study of the behavior of these animals in the wild. An understanding of natural communication among chimpanzees is also important as background for the laboratory studies we look at in Chapter 10.

Chimpanzees seem to have a vocal repertoire roughly comparable to that of vervets, but it does not include a system of alarm calls differentiated in terms of the predator. Basically, only one vocalization indicates the presence of predators such as leopards, pythons, or human hunters: the *wraaa*. This lack of differentiation is surely connected with the fact that chimpanzees are larger, less vulnerable animals than vervets (or ring-tailed lemurs), and basically have little to fear in their environment (or at least, that was the case prior to the arrival of humans). When danger does appear, chimpanzees tend to respond in a uniform way: by climbing up into the trees. There is no particular pressure on chimpanzees to develop a set of finely differentiated alarm calls.

Another area where we might look for an attribution of reference in chimpanzee vocalization is food calls. Chimpanzees give a *food grunt* or a *food aaa* call when they find a source of food, and that attracts the others in the neighborhood to come and share the find. They are likely to call more extensively when the quantity of food discovered is greater. A number of different kinds of food might be found in the wild, with different ways of gaining access to it, so we might imagine that different calls would be made for different foods, but nothing of the kind seems to happen.

The one place where we can be sure that variations on a single theme make a difference is the chimpanzee's famous *pant-hoot*. This call is given by animals traveling or nesting alone, or between nesting families at night. One chimpanzee makes a series of pant-hoots and waits. Others respond in kind, and they alternate their calls. Since the pant-hoot of each individual is distinguishable from that of the others (presumably to the chimpanzees as well as to human observers), it serves to identify animals who may or may not be in sight of one another.

Pant-hoots are used too to identify oneself to one's fellows. We might almost think of them as individual "names," except that no chimpanzee ever produces any other's pant-hoot. That is, while we can think of Panzee's pant-hoot as something like an announcement "I'm here!" no one ever asks "Is that you, Panzee?" or "Where's Panzee?" Nor does anyone try to deceive the others by claiming falsely to be Panzee. While chimpanzees are certainly aware of the identities of the members of their group and others, and attend to differences in pant-hoot calls for information about the individuals around them, we have no reason to think of these calls as referring to individuals in any more interesting sense.

Chimpanzees make a number of other vocal calls to each other within groups. Some are the greeting of a subordinate to a dominant member of the group (*pant-grunt*) and, conversely, from the dominant to a subordinate (*soft bark*). At least three different sorts of *scream* are given by animals who are attacked, or upset, or copulating. The acoustic differences among these screams are not well understood, but other animals react differently to them, so there must be a distinction. A few miscellaneous calls of distress or surprise round out the inventory.

Chimpanzee vocalizations may be under the control of the animal's limbic system, rather than the cortical regions we associate with voluntary activity, although (as in the case of most monkeys) this is an issue that needs further research. Calling is by no means entirely automatic and involuntary. Numerous observations attest that chimpanzees can suppress their calls. They walk very quietly together when hunting other animals, for example, and a female suppresses her copulation screams when she is with someone other than the dominant male (who would probably respond by attacking her partner if he became aware of what was going on).

Other observations show chimpanzees can behave in ways that seem deceptive. They may give alarm calls when no danger is present so as to get another animal to go away, or they may not give food calls when there is a likelihood that such a call would cause others to come and take all the food. The former possibility is particularly important to explore further, as Goodall says that "the production of a sound in the *absence* of the appropriate emotional state seems to be an almost impossible task for a chimpanzee." By and large, their calls are closely bound to, and expressive of, these internal emotional states, rather than referring to things outside of themselves.

Gestures

Calls are not the only way (or perhaps even the most important way) chimpanzees communicate with one another. Many of their gestures are expressive too—apparently intentionally so. Goodall reports more than a dozen distinct gestures that the animals she observed use for social purposes. Probably there are many more, as confirmed by other observers of animals both in the wild and in captivity. Several of the signals are used to initiate play, while others solicit food, grooming, or other personal attention.

These gestures are more likely to be under conscious control and not automatic. The vocal signals discussed above seem to be largely an expression of the animal's internal state, and the information they convey to others is more or less a by-product of that. Manual gestures, in contrast, are apparently intended (in at least some cases) to produce a particular effect in others. Chimpanzees can be seen to make a gesture, and then to pay close visual attention to the other to see if the right reaction is forthcoming; they often repeat the gesture more emphatically if it is not.

Some, but not all, of these gestures are apparently quite natural and could be argued to be innate. In some cases, animals have obviously made up new gestures on the basis of novel circumstances. For instance, throwing wood chips at another chimpanzee to initiate play is unlikely to be an innately specified behavioral routine. An individual was observed to devise a strategy of passing by another while offering up one limp leg, inviting the other animal to grab it and play—all the while alternating his gaze between the potential playmate and the leg. Chimpanzee gestures seemingly are flexible in formation and use.

Not only manual gestures are used in this way. While chimpanzees' facial expressions are nowhere near as controllable as are those of humans, some — such as a fear grimace — are characteristic. Anecdotal reports tell of animals concealing their faces with their hands so as (apparently) not to reveal their fear. One gorilla covered its "play face" with its hand, presumably so as not to give away its playful intention. Goodall reports a particular gait that male chimpanzees use when leaving camp, a way of walking that conveys determination of purpose and invites others to follow. Reported instances of apparently deceptive chimpanzee behavior involve the use of this special gait to induce others to go off in one direction while the animal himself goes elsewhere.

Finally, there is evidence for the use of objects as symbols. Chimpanzees in some groups are observed to trim and groom leaves in the presence of a desired female, as an invitation to mate. And bonobos traveling in the forest along networks of trails have been observed to stamp down vegetation or leave large leaves pointing in the direction they have taken at forks in the road, from which others in their group can follow them.

Except in the limited sense that all are means of communication, these gestures have little in common with human language. But the fact that chimpanzees and bonobos do make flexible use of gesture in the wild provides an important bit of context for the laboratory studies we will consider in Chapter 10. That nonhuman primates can be trained to use manual gestures in a meaningful way is to some extent a natural extension of normal behavior.

This is not to deny that it *is* an extension, especially to the extent that we can see the meanings of their learned gestures as arbitrary, and not fully iconic. At least some of the most widely reported "signs" that turn up in the vocabularies of language-trained chimpanzees and bonobos (such as COME-GIMME and MORE) turn out to deviate markedly from the American Sign Language signs that are their presumed models. Significantly, the deviations are in the direction of naturally occurring manual gestures with very similar content, documented by Goodall and others for animals in the wild. The question arises of whether those signs actually form part of a language-like system being learned, or whether they are simply adaptations of something quite different. To understand that issue, we need to appreciate something of the richness of the animals' natural gestural communication.

8

Syntax

"Get your own breakfast first, Stubbins. I'll take another little sleep now. And don't forget the cat, will you?"

"No, doctor," I said, "I won't forget."

"By the way," he added as I pulled aside the tent flap to leave, "you will find her difficult to talk to. Took me quite a while to get on to the language. Quite different from anything we've tried so far in animal languages. A curious tongue—very subtle, precise and exact. Sounds as though whoever invented it was more anxious to keep things to himself than to hand them over to others. Not chatty at all. There's no word for *gossip* in it. Not much use for people who want to be chummy. Good language for lawyers, though."

—*Doctor Dolittle's Return*

We have now looked at the communication systems of a number of non-human species. In essentially every case, I have raised the point that no matter how fascinating the behaviors involved, they still only allow the animals that use them to convey messages from a small, fixed set.

Honeybee waggle dances look like an exception, but they escape this limitation only on a technicality. Distinct dances could in principle refer to food sources at an unlimited number of locations, but that is all the bees can "talk about": the location of food (or a potential new hive site). And the theoretical precision of this system is not actually realized: bees cannot reliably divide the scales of their dances very finely. One estimate suggests that the dance contains only about four bits of information with respect to direction, for instance, so that the 360 degrees of the compass are effectively divided into no more than a couple of dozen distinct values.

Human languages contrast with every other system about which we have any serious knowledge. The range of things that we can express (and that our listeners can understand) in any natural language is unlimited. What is it that gives language this special expressive power?

Part of the answer is the richness of our vocabulary. Let us accept that the number of different words any particular individual knows is finite. This statement, seemingly obvious, in fact is probably wrong. Every language has ways of making new words on the basis of existing ones—through compounding and derivational morphology, for instance. Anyone who knows what a *puritan* is and also knows about the general properties of words beginning with *anti-* (such as *anticommunist*) "knows" the word *anti-puritan* without ever hearing or using it. In some languages, such as those of the Inuit, such morphological construction of words from other words is used much more exuberantly than in English, and the essential unbound-edness of the set of words becomes apparent.

An average college student probably has a vocabulary of about fifty thousand words. Such a number vastly exceeds the message repertoire of any known nonhuman species. But surely the story cannot end there. However many words we know, the limited size of that set does not stop us from being able to talk about far more than fifty thousand different subjects (or indeed any particular number) without having to learn or to coin new words. Even if we did not have the word *gossip*, we could still talk about gossiping by describing it with combinations of other words. Cultures in which modern kitchen appliances are unknown may eventually borrow a word such as *refrigerator;* but until they do, they can get along nicely by talking about "the box you put things in to keep them cold." This open-ended character of language comes from the possibility of systematically combining the words we know into an unbounded range of sentences — the system of *syntax*.

The word *syntax* comes from Greek *sun+tassein* "to arrange together," and the way we put words together is what we want to understand. If combining meaningful symbols is all that is involved in producing a system with the expressive power of a human language, it stands to reason that many animals ought to be capable of doing that, on some scale. In a few cases, including those of the Diana and Campbell's monkeys, there is evidence that a combination of signals can convey a message that is not the same as the message conveyed by any of the individual signals. Why, then, are we sigling out "syntax" for such special attention?

The phrases and sentences of natural language are not just groups of words joined together. They have an intricate internal structure (which our high school English teachers try to make us analyze), and this structure is essential to their efficacy in expressing complex meanings. Grammatical organization may seem like something invented and taught in school, learned painfully and late in life. Yet we find the same basic structural properties in every human language, including those that are neither written nor explicitly taught. These characteristics are part of what makes a language a language, and they are quite distinct from the explicit prescriptions that constitute so much of our education about "grammar."

Not only is there a structure to the sequences of words that make up expressions in natural languages, there is also a more abstract, "underlying" structure that is distinct from (but systematically related to) their surface form. Most surprising of all, perhaps, the syntactic system of a natural language has essential properties that are all its own, which do not follow in any serious sense from those of the meanings we convey or the modality in which we convey them.

In the discussion that follows, the technical details that arise may seem arcane, and remote from any interest in animals or the general nature of communication systems. The study of syntax is highly technical, and the modern literature of the field is extremely complicated. Much of that complexity does not bear on the general themes of this book, but some of it does. Without an understanding of the type of intricate structure that characterizes every single human language, we cannot possibly appreciate just what makes language what it is, and what is truly involved in comparing it with other forms of communication. It is precisely by virtue of these syntactic mechanisms that natural languages provide us with a unique, wonderful, and infinitely flexible tool.

Syntax Is More Than Combining Words

Suppose our language did not have any syntax, and that what we know really was just a collection of words. Lots of them perhaps, but still a finite collection. That would mean we were able to talk only about a fixed (if large) range of things—namely, what we happen to have words for. If we did not have a specific word for "the third person in from the aisle in the front row of the upper deck," we could not refer to that specific individual. Of course, if we were actually at the ballpark, we might be able to say "Person!" and point to the one we meant. But even without special words, given the resources of English I am able to tell you that "the third person in from the aisle in the front row of the upper deck caught Bonds's home run ball, but the guy behind him grabbed it away from him." You can understand that without either of us, or the individuals referred to, being present.

What gives us the power to talk about an unlimited range of things even though we only know a fixed set of words at any one time is our capacity for putting those words together into larger structures, whose meaning is *compositional.* That is, the meaning of a combination of words is in general a function of the meaning of the individual words, along with the way they are put together. Compositionality means that we do not have to learn all the meanings we might need individually, since we can make up new expressions of arbitrary complexity by putting together known pieces in regular ways.

Furthermore, the system of combination is *recursive.* That is, we only need to know how to construct a limited number of different kinds of structures, because those same structures are reused as building blocks. Sentences contain other sentences as pieces (sentence complements, for example, or relative clauses). The property of recursion allows us to build structures that are unlimited in complexity, while making use of only a limited range of basic patterns. Language thus makes infinite use of finite means through what Pinker calls a *discrete combinatorial system.*

Of all known communication systems, these properties are unique to human language. Virtually every animal has some way to communicate with others of the same species, but apart from human language, it is limited to a rather small, fixed set of "words." This fixed set consists of unitary gestures, sounds, chemical or electrical signals, each with a set meaning, where the only variation in that meaning comes from the situational context in which they are produced.

Vervet monkey alarm calls distinguish among a small number of different predators (leopard, eagle, and snake). Some groups have adapted other calls to permit them to indicate something like an "unfamiliar human," while others have similarly developed a call for dogs accompanying human hunters. This example shows us that, over time, the monkeys can apparently add to their vocabulary, unless what is going on is simply that they are developing a new use for a vocalization that is part of their innate stock. The system appears to be largely innate, and the mechanism involved remains somewhat ambiguous. Nonetheless, any such augmentation can occur only slowly, and the system itself still remains small. If little green men with a distinctive hunting style were to descend from flying saucers into the vervets' world, the monkeys would be unable to describe the new danger to one another. Vervets do not even have a way of saying anything about "the leopard that almost sneaked up on us yesterday."

English must be something more than the sum of its words. But how are we to describe the "something more" that is needed? In other words, how can we characterize what it is to be a sentence of English? A first guess might be that "an English sentence is a sequence of English words," but that theory is much too weak, and also too strong. On the one hand, some English sentences may contain things that are not, in fact, English words (*Why does my cat go "Grrr!" when I try to pet her?* or *Fred stopped singing "Auprès de ma blonde" immediately when the children came in*). On the other hand, some strings of English words are not sentences (** Cat mat on the is the*). Remember our convention of preceding an ungrammatical string of words with an asterisk.

Ignoring the oddities that can appear inside quotations, we might try to amend our simple theory to say that "an English sentence is a *meaningful* sequence of English words." Quite apart from the question of how we might determine meaning, this theory does not fare much better. As we noted in Chapter 3, some strings of words that do not really have a coherent meaning are nonetheless grammatical sentences of English in ways that similar, equally incoherent strings are not. Recall Chomsky's famous pair in (1):

(1) a. Colorless green ideas sleep furiously.

    b. *Furiously green sleep ideas colorless.

The other side of this coin consists in those strings of words to which we *can* assign a coherent meaning, even though they violate the words of the language. Much poetry, in fact, plays on this fact directly.

Another obvious nonstarter of a theory is the notion that you have learned all the sentences in advance and can pick them out by consulting a list. That would just be a variant of the theory that claimed no syntax at all. Such a list would have to be finite (our brain, where we would have to be storing the list, is after all finite, and so is our experience of sentences to date), but the range of sentences we can produce and understand is not.

We might think we could combine these notions to get a better theory. Take the words of our language and a limited set of sentences, and say that other sentences are merely put together on the basis of sentences like them. But what does it mean for a new sentence to be "like" an old one? Is this any different from saying that new ones are made "by analogy," a notion that does not really help with problems like these? If the basic mechanism of linguistic creativity is the same as that which allows students to perform well on standardized tests with their substantial "analogy" sections, why is it that given *ear* and *hear*, together with *eye*, we cannot create a verb *heye* meaning "see"?

For a more syntactic example of this point, notice that "yes or no" questions are systematically related to simple sentences on the basis of a very specific pattern—a pattern that is only one of many possible "analogies" one might make to declarative sentences.

(2) a. Fred is really a werewolf.
    b. Is Fred really a werewolf?
    c. The guy who is showing his teeth is really a werewolf.
    d. *Guy the who is showing his teeth is really a werewolf?
    e. *Is the guy who showing his teeth is really a werewolf?
    f. Is the guy who is showing his teeth really a werewolf?

Why is it impossible to form a question by inverting the first two words of a declarative, as in example (2d), or by moving the first auxiliary verb to the front as in (2e)? Either of these patterns could be based on the relationship in simple cases such as (2a) and (2b), and either would be simpler than the formally more complicated relation that results in (2f). If a general notion of analogy is at work here, one or both of these patterns ought to support valid analogical formations.

But they do not, and indeed no one needs to be told to avoid forming questions after the manner of (2d) and (2e). Some analogies work and others do not. The ones that work we can call "rules of grammar," and when we explore this notion further, it turns out to have such a specific sense that there is no advantage to be gained from thinking of what is going on as a matter of "analogies." In the case of syntactic regularities, for instance, it seems that by and large only "analogies" that are based on treating whole phrases as units are even candidates for consideration. This limitation follows from the specifics of syntactic organization, not from a general concept of analogy.

Words and Phrases

We need to give an explicit account of what the regularities are that characterize English sentences. A simple way of doing so might be to develop a model that tells us which sequences of words are sentences. Given a word that is part of a sentence, the rules in such a model would tell us which other words could follow it. For instance, following the word *I* a sentence might continue with words such as *like, eat, sleep* but not with *the* or *blue.*

It is well known, however, that such a *finite state device* cannot possibly work, because in essentially every language we find cases of dependencies between a word or phrase that appears at one point and something else that may not show up until much later in the sentence. Because the number of words that can intervene between the parts of such an *unbounded dependency* is in principle unlimited, it is impossible to specify grammaticality in terms of a strictly local, word-by-word notion.

The way our knowledge of grammar overcomes this problem is by treating sentences as made up not just of strings of words, but also of larger, meaningful, multiword phrases. Regarding the structure of sentences in terms of their constituent phrases goes a long way toward solving the problem posed by unbounded dependencies. Consider the sentences in (3):

(3) a. Fred bequeathed Fido to his son.
   b. Fred bequeathed an old spotted dog with one lame leg to his son.
   c. Fred bequeathed a parrot that had bitten the old spotted dog that tried to knock its cage over by leaping up from the sofa to his son.

The verb *bequeath* requires that the sentence in which it appears specify the recipient of the bequest. We cannot just "bequeath" something; we must bequeath it to someone. In all of the sentences in (3), the phrase *to his son* fills this role, but it would be difficult to be sure of that by looking at what comes right after *bequeath*.

From the point of view of the structure of the sentence, though, all of the expressions *Fido, an old spotted dog with one lame leg,* and *a parrot that . . . up from the sofa* play the same role: they identify the bequest. In these structural terms, then, the phrase *to his son* is separated from the verb by only one phrasal unit. That unit may be relatively short (*Fido*) or as long as the speaker has the patience to make it, but it is still a single unit in terms of the sentence's phrase structure. A grammatical description based on phrases (rather than words) as units can give a much more satisfactory account of the grammar of sentences in natural languages.

The same string of words may have more than one organization into phrases. Consider the sentence *Jones shot the guy with a gun in his hand.* That could mean *It was the guy with a gun in his hand whom Jones shot,* in which case *the guy with a gun in his hand* is being treated as a unitary phrase—a fact we can confirm by looking at the passive sentence *The guy with a gun in his hand was shot by Jones.*

This same sentence could also mean *It was with a gun in his hand that Jones shot the guy,* and here the organization into phrases is somewhat different. We can again confirm this interpretation by considering the passive: *The guy was shot by Jones with a gun in his hand.* Or else the exact same sentence could mean *It was with a gun that Jones shot the guy in his hand.* And there is yet another sense: *It was the guy with a gun that Jones shot in his hand,* whose passive is *The guy with a gun was shot by Jones in his hand.*

There are thus at least four distinct senses of the very same sequence of words, yielding a four-way ambiguity. Either, both, or neither of the expressions *with a gun* and *in his hand* can be associated either with *Jones* or with *the guy.*

Many English expressions are ambiguous. *Jones gave Smith an unreasonably long sentence* has different meanings depending on whether Jones is a judge or an English teacher. The difference hinges on the ambiguity of the word *sentence.* But in our sentences about *Jones* and *the guy,* there are no ambiguous words, at least none whose ambiguity seems related to the range of meanings we find. Rather, the different senses result from *structural ambiguity,* which is part of what we know about English sentences. In order to express such facts, any account of our syntactic knowledge has to embody a description of sentences as structures and not just as strings of words.

How do we know what the organization of a sentence into phrases is? Consider the argument just given about how to disentangle the *guy with a gun* sentences. Often two or more constructions are systematically related to one another, such that the same material appears in different places in the two. For instance, a passive sentence like *The petition was denied by the court* is systematically related to the active sentence *The court denied the petition.* The direct object of the active sentence appears as the subject of the passive, and the subject of the active appears in a phrase with the preposition *by* (in addition to some changes in the form of the verb).

A generalization emerges when we examine relations like this. It is complete phrases whose position can be altered when we construct the passive corresponding to a given active sentence, not individual words. The same is true of the construction relating *Fred fed the cat* to *It was Fred who fed the cat,* and many others. The same sequence of words making up a complete phrase appears in different positions in the two related constructions, while retaining the same relation to the overall interpretation. The number of words involved is irrelevant: what matters is that together they constitute a phrase. This phrase may itself be part of some larger phrase(s), of course, but its own character as a syntactic unit is what counts.

For another example, consider the way we form questions. When we question something, we must in general question an entire phrase:

(4) a. Jones is in love with a student of feline genetics.

    b. Which student of feline genetics is Jones in love with?

    c. *Which student is Jones in love with of feline genetics?

It is whole constituents that are affected when we form questions. We can thus conclude from facts like those in (4) that the phrase structure of *a student of feline genetics* is [[a] [[student] [of [feline genetics]]]]. In this structure, *a student* is not a phrase, so it is not eligible to be questioned by itself. That accounts for the ungrammaticality of sentence (4c).

Apart from resolving ambiguities, such tests can also be used to address the fact that very similar sequences may have to be assigned different analyses into constituent phrases. Consider the sentences in (5):

(5) a. Pat ridiculed [Harry's theory that the moon is made of green cheese].

    b. Pat persuaded [Harry's students that the moon is made of green cheese].

The expressions in brackets in these sentences are parallel if we look at them as strings of words, but are they both phrases? If so, how can we tell? Consider what happens when we form passives.

(6) a. Harry's theory that the moon is made of green cheese was ridiculed by Pat.
   b. *Harry's students that the moon is made of green cheese was/were persuaded by Pat.
   c. Harry's students were persuaded by Pat that the moon is made of green cheese.

We see that in the one case, the result is acceptable, while in the other case it is not. On our assumption that it is *phrases* that can constitute structural units such as the subjects of passive sentences, we get different answers about the "phrase-hood" of the bracketed strings: *Harry's theory that the moon is made of green cheese* seems to be a phrase, while *Harry's students that the moon is made of green cheese* seems not to be. In contrast, *Harry's students* clearly *is* a phrase. Taken together, these considerations suggest quite different structures for the two sentences in (5).

To confirm this result, let us apply the question test to the sentences in (5):

(7) a. Whose theory that the moon is made of green cheese was ridiculed by Pat?
   b. *Whose theory was ridiculed by Pat that the moon is made of green cheese?
   c. *Whose students that the moon is made of green cheese were persuaded by Pat?
   d. Whose students were persuaded by Pat that the moon is made of green cheese?

Again, if we assume that it is *phrases* that can be questioned, we get different results in these examples for the sequences *Harry's theory that the moon is made of green cheese* and *Harry's students that the moon is made of green cheese*. The first seems to be a phrase (while the included part of it made up of *Harry's theory* is not here a phrase by itself); whereas *Harry's students that the moon is made of green cheese* is not a phrase (although the included sequence *Harry's students* is). The structures of the sentences in (5) thus seem to be as in (8):

(8) a. Pat ridiculed [Harry's [theory [that the moon is made of green cheese]]].

b. Pat persuaded [Harry's [students]] [that the moon is made of green cheese].

Phraʃal Typeʃ

As speakers of a language, we generally have fairly clear intuitions about what sequences of words constitute phrases, and we can usually find some confirmatory evidence, but the evidence is not always of the same type. That in turn suggests that phrases are not all the same kind of thing: that is, there are different kinds of phrase. Another test elaborates the point. We can often replace a phrase with a *pro-form* of some kind, if its meaning is determined by the context. When we do that, different sorts of phrases require different pro-forms.

(9) *Pronoun:*

a. I met (Fred / a lady with a feathered hat / the tall blond man with one black shoe about whom you had warned me).

b. I met <u>him</u>.

Proadverb:

a. I met Fred (in the subway / on top of Mount Catoctin / in the middle of the square where they sell those funny little wooden dolls).

b. I met Fred <u>there</u>.

Pro-verb:

a. I (met Fred / told Fred that those unfiltered cigarettes will kill him) and Melissa (met him / told him that) too.

b. I told Fred that those unfiltered cigarettes will kill him, and Melissa <u>did</u> too.

We want to categorize phrases in two principal ways. On the one hand, we want to make a distinction that corresponds largely to the differences among pro-forms. Phrases that can be substituted by a pronoun and those that can be substituted by *do* and similar words belong to different types in this sense. The difference depends on the internal structure of the phrase — we can say that every phrase is built up around some central element that constitutes its "head." Grossly, the head is an element of a part-of-speech class that can function all by itself in the same way as a phrase constructed around it. When a phrase is headed by a verb, it is a verb phrase; when it is headed by a noun, it is a noun phrase, and so on. Actually, some theories of syntax consider that the real head of what we are calling a noun phrase is the determiner (*three, a*) rather than the noun, and call it a determiner phrase. We will ignore that refinement and stick to the more traditional view.

(**10**) *Noun Phrase (NP):*

    a. I saw ([$_{NP}$ three brilliant purple [$_N$ finches]] / [$_{NP}$ a whole [$_N$ bunch] of birds with long feathers] / [$_{NP}$ a [$_N$ bird] that was eating a fish]).

    b. I saw [$_{NP}$ [$_N$ birds]].

Verb Phrase (VP):

    a. Last night I ([$_{VP}$ [$_V$ ate] a plate of spaghetti] / [$_{VP}$ [$_V$ looked] into the mirror] / [$_{VP}$ [$_V$ gave] Louise my phone number] / [$_{VP}$ [$_V$ did] what you told me to do]).

    b. Last night I [$_{VP}$ [$_V$ slept]].

Prepositional Phrase (PP):

    a. I walked ([$_{PP}$ [$_P$ into] the middle of an argument] / [$_{PP}$ [$_P$ over] the yellow line]).

    b. I walked right [$_{PP}$ [$_P$ in]].

These are matters of the way in which a given phrase is constructed out of words and other phrases of specific types.

Phrase Structure and Grammar

We also want to distinguish phrases in terms of the way they relate to the rest of the sentence in which they appear. For instance, a noun phrase — the very same noun phrase — can be used as the subject S of a sentence, or as a direct object, or as an object of a preposition, and in other ways as well.

(**11**) a. My obnoxious cat hisses at me.

    b. I don't understand my obnoxious cat.

    c. This is a story about my obnoxious cat.

    d. My obnoxious cat's behavior is indefensible.

Our knowledge of phrase structure thus has two distinct components: first, how phrases of various sorts (NPs, VPs, and the like) are constructed; and second, how they are used. These are not independent of each other, however. The ways in which a given phrasal type can be used are precisely the ways in which that type can contribute to the construction of *other* phrases. NPs in different positions within the structure of a sentence or a

VP serve as different *arguments* (subject, direct or indirect object, object of a preposition, and so on) within that structure. Considered as NPs, though, they all have the same possibilities of internal composition, regardless of the role they play in the larger structures of which they are components.

We can formalize the two together, saying for instance that (in English) a sentence can be composed of a noun phrase followed by a verb phrase. The verb phrase, in turn, can be composed of a verb alone, or a verb followed by a noun phrase, all (optionally) followed by a prepositional phrase. Noun phrases, in their turn, can be composed of a determiner (a number, an article, a possessor phrase) followed (optionally) by a smaller phrase that we will call an N' and that can consist of an (optional) adjective followed by the noun. Prepositional phrases, in turn, are made up of a preposition followed by a NP.

(12) a. S → NP VP

    b. VP → V (NP) (PP)

    c. NP → Det (N′)

    d. N′ → (Adj) N (PP)

    e. PP → P NP

Many patterns in English sentences are not described by (12). In order to cover anything like all of the language, we would need to expand this set of statements substantially. Still, the extensions that would be required are not limitless, as the range of sentences covered by the overall system of rules must be. Looking at the small set of rules in (12), we can see how the property of recursion contributes to this result. NPs usually contain a phrase of type N′, and this in turn can contain a PP—which itself will contain another NP. With only a small set of rules, we can describe phrases such as *the cat with a scratch on the third finger of the right front paw* without limit.

Recursion contributes mightily to the task of allowing a finite system to express an unbounded range of meanings, and without it the combination of symbols into groups remains extremely limited. A Diana monkey may understand the sequence of boom call plus leopard call from a nearby Campbell's monkey as indicating "Maybe a leopard," a different message from "Leopard!" The combination appears to follow a regular principle, but one that is quite restricted. Each component of the combinations that we find in natural primate signaling systems must come from a short, fixed set, and the result is not substantively different from adding a few more basic calls to the list of possibilities. Only when combination becomes hierarchical and recursive does it escape from such limits and achieve the openendedness of human language.

Another point about the rules in (12) is that they define both of the aspects of phrases that we have just been discussing. Being a noun phrase is defined here as having the structure defined by rule (12c). Being a subject, on the other hand, can be defined as being the NP in rule (12a), and so forth.

When we give a *particular* set of rules like those in (12), we describe a particular set of possible structures. For instance, English puts the verb before its direct object, as described by rule (12b). Other languages, in contrast, put the verb after its object, as in Japanese or Turkish. Those languages would not be described by (12b), but rather by a slightly changed version of that rule: VP → (NP) V.

Phrase structure provides us with a rich framework for describing sentences. We might propose a theory of syntax that consisted in specifying the structure of sentences, in terms of phrases, by a system of *phrase structure rules* like those in (12). This theory would reconstruct our knowledge of what is (and is not) a sentence in our language by saying that exactly those strings of words which can be analyzed in terms of these rules is a sentence. In contrast, a string of English words that does not correspond to any phrase structure provided by those rules is ipso facto not a sentence. We learn in the following paragraphs that this definition does not suffice, but it is far better than the "string of words" theory.

Syntax seen this way is an autonomous component of grammar. That is, it does not depend on semantics, the meaning of sentences. Rules of grammar like those in (12) describe sentence structures, per se, on the assumption that other principles exist to tell us how to link them up with meanings. With rules like (12), we can describe *Colorless green ideas sleep furiously* just as well as *The cat is on the mat*. Syntax describes a range of structures, and these can be related to meaning on the one hand and to sound on the other hand, each of which is described in its own terms. Ideally, a well-formed sentence has a well-formed interpretation and pronunciation, but that is not necessary for us to recognize Chomsky's *Colorless* . . . sentence as grammatical (though meaningless).

At first glance, this suggestion is rather surprising. If we think of languages as systems for associating sound with meaning, as they are often portrayed, it would seem that semantics (or perhaps phonology) ought to play the central role in defining the class of sentences that make up a language. The set of syntactically structured sentences might be expected to derive from the class of meanings they are expected to express. Syntax itself would seem to be a pretty arbitrary part of language—maybe even marginal.

Nonetheless, the syntactic organization of sentences cannot be merely a property that derives from their meaning, since essentially the same meaning can be conveyed by sentences of different syntactic structure.

(13) a. My cat worships the food bowl.
    b. My cat is devoted to the food bowl.
    c. I figure that the cat has gone out.
    d. I am of the opinion that the cat has gone out.

Quite generally, the behavior of sentences (that is, their relation to other sentences) depends on their syntactic form rather than on their meaning. Different sentences that have the same meaning can behave quite differently syntactically when they differ in syntactic form.

(14) a. The food bowl is worshiped by my cat.
    b. *The food bowl is been devoted to by my cat.
    c. I figure the cat to have gone out.
    d. *I am of the opinion the cat to have gone out.

The structure we assign to sentences is crucial to the way we understand them. Recall our discussion in Chapter 3 of the interpretation of pronouns and consider the sentences in (15):

(15) a. Chelsea thought that she should go to Yale.
    b. She thought that Chelsea should go to Yale.
    c. That Chelsea was going to Yale excited her.
    d. That she was going to Yale excited Chelsea.
    e. Her mother thought that Chelsea should go to Yale.

In all the sentences in (15), the pronoun (*she* or *her*, depending on the grammatical position) may be interpreted as referring to some other female person, not Chelsea (or her mother). It is also possible to interpret the pronoun as referring to Chelsea—*except* in sentence (15b), where *she* must refer to Chelsea's mother, or to a girlfriend, or in fact to any female human or pet animal in the universe, except Chelsea. Why should that be?

One possible factor is that in (15b) the pronoun *she* occurs earlier in the sentence than *Chelsea,* and it might seem plausible that a pronoun could not come before another NP that supplies its actual reference. That cannot be what is going on, though, because in sentences (15d) and (15e) the pronouns *she* and *her* precede *Chelsea* and it is still possible for them to refer to her—unlike sentence (15b).

The principle that governs this surprising difference is based solidly on the phrase structure of the sentences in (15). The problem in (15b) is that the pronoun not only appears before another mention of Chelsea in the same sentence, but also "higher up" in the structure. In (15b) *she* is the subject of the verb *thought,* while the later occurrence of *Chelsea* is within an embedded sentence that serves as the direct object complement of *thought.* The pronoun is thus structurally superior to the full NP *Chelsea,* and under these specific circumstances (whose definition is purely structural) the two cannot refer to the same individual.

To make this account complete, we need a precise definition of "higher up." In sentence (15e) the pronoun *her* precedes *Chelsea,* but since it is not the subject (it is only embedded within the subject, in a possessive relation), it does not count as higher up, and the definition must reflect this relationship.

Remarkable as these facts may seem once they are pointed out, they are not at all isolated peculiarities of English. Indeed, every language investigated to date displays essentially the same conditions on pronouns. In particular, a pronoun that appears both preceding and higher up in the structure than another NP cannot refer to that same individual.

These matters may seem to fall somewhere on the margins of language, but that is surely an illusion. The conditions of interpretation are precise, and they recur across human languages. They could not possibly have been learned explicitly (ask yourself how you came to know how to interpret the pronouns in our examples!), and it is plausible to suggest that they are part of what it is for a system to *be* a human language.

These principles depend in their essence on the notion that sentences have a finely articulated internal structure in terms of phrases, rather than being just sequential combinations of words. Linear order is not enough by itself to describe the possibilities of reference for pronouns—we need to know about the positions of words and phrases within a fully articulated phrase-structure representation of sentences.

Abstract Syntactic Structure

Sentences are more than strings of words, then, and the structure within which those words appear is important in determining which strings are grammatical in a language and what they mean. But syntactic structure in natural languages is even richer than the phrase structure representations we have just been discussing. In fact, in order to give a complete account of a given sentence, we need to supply more than just the organization into phrases. Some of the additional information that plays a role is abstract and only indirectly related to the sentence's surface form. To see this, we return to a class of sentences that came up earlier.

In our discussion of the tests that can confirm our intuitions about phrase structure, one construction was the formation of "content" or "information" questions. The set of phrase structure rules in (12) does not provide any way to describe such sentences, though. Compare the structure of a content question with that of a corresponding declarative.

(16) a. What will Pedro win this year?
   b. Pedro will win the Cy Young award this year.

These sentences contain an auxiliary verb (*will*), for which our rules do not provide. The analysis of auxiliaries is complicated. Most syntacticians would say that the great success of Chomsky's proposals for syntactic analysis was (at least initially) due to the satisfying account he offered in his 1957 book, *Syntactic Structures*, for problems in this area that had confounded generations of linguists. The analysis of auxiliaries is not really germane to the present discussion, however. Let us simply assume that they can be provided for, and focus on the possibility that we might account for content questions by adding the following rule to the system proposed in (12).

(17) S → Question-Phrase Auxiliary NP VP

Obviously a lot of details need to be taken care of, but let me draw attention to one particular point. A rule like (17) raises new problems for the statement of generalizations that otherwise seem straightforward. Remember the verb *bequeath:* when this verb is used in a sentence, it must be accompanied by mention within the VP of both the bequest and the legatee. **Fred bequeathed the family home* and **Fred bequeathed to his cat* are incomplete on their own, as opposed to *Fred bequeathed the family home to his cat.* It is entirely typical for verbs to be associated in this way with specific requirements on what kinds of arguments accompany them. In *The Language Instinct,* Pinker describes this by saying that "within a phrase . . . , the verb is a little despot." The class of verbs called intransitive cannot have both a subject and a direct object, whereas transitive verbs can (and sometimes must) have both.

(18) a. I think Fred is sleeping.
   b. *I think Fred slept his cat.
   c. Fred's cat ate the linguini.
   d. Fred's cat is eating. [≈Fred's cat is eating something.]
   e. Fred owns a cat.
   f. *Fred owns.
   g. Fred put his cat in the closet.
   h. *Fred put in the closet.
   i. *Fred put his cat.

The sentences in (18) illustrate a number of facts about the verbs that appear in them:

(19) a. *sleep* appears only when no NP follows within the VP (that is, *sleep* is intransitive).
   b. *eat* appears with a following NP; when none is present, an indefinite is assumed (*eat* is optionally transitive).
   c. *own* requires a following NP (*own* is obligatorily transitive).
   d. *put* requires <u>both</u> a following object NP and a following locational expression.

Let us call generalizations like these the argument frame requirements of a verb.

The sentences in (18) illustrate the argument frame requirements of various verbs. Now compare them with the corresponding content questions, in (20). It looks as if the argument frame requirements of verbs are systematically different when they appear in content questions as opposed to simple sentences.

(20) a. Who do you think is sleeping?
　　b. *Who do you think Fred is sleeping?
　　c. *What is Fred's cat eating the linguini?
　　d. What kind of cat does Fred own?
　　e. *What kind of cat does Fred own a Maine Coon?
　　f. Whose cat did Fred put in the closet?
　　g. In which of the closets has Fred put Pooh now?
　　h. *What did Fred put Pooh in the closet?

At first glance, some of these content question sentences seem to be missing one of the NPs required by a verb (perhaps the subject, perhaps one of those that belongs inside the VP). In others, when the VP has exactly the number of NPs that it "ought" to have, the sentence still feels as if it contains something extra, something that cannot be digested by the structure of the sentence.

The real generalization, of course, is that the question phrase at the beginning of these questions provides an argument that is required somewhere within the sentence, even though it is not in the position where the corresponding phrase would be expected. In traditional grammar, we might say that "the question phrase is *understood as* the subject, direct object, and the like."

How can we make sense of this notion of the question phrase as being "understood as" filling some particular argument function? Phrase structure rules cannot express this concept, since they merely tell tell us how the words and phrases that we have before us are organized into larger phrases. To correct for this limitation, syntacticians assume that no single set of phrase structure rules describes the complete structure of sentences by itself. A comprehensive description involves (at least) two separate components, each of which accounts for part of what we need to say about sentence structure.

To see how this works, let us return to the analysis of content questions. These obviously bear a systematic relation to the corresponding declaratives: the question phrase can be regarded as substituting for some constituent phrase in the declarative sentence, although it appears at the front of the sentence instead of where an argument phrase of its type would be expected. The sentence in (21a), for instance, is "understood as" something like (21b):

(21) a. What might Robin believe that Lee found in the attic?
    b. Robin might believe that [Lee found [what] in the attic].

We could say that the structure of (21a) is really something like (22). The structural position in which we would expect to find the direct object is actually empty, as represented by the "[e]." That is, the sentence as pronounced has nothing in the direct object position, although an element that is pronounced elsewhere is interpreted as if it were there.

(22) [what] might Robin believe that [Lee found [e] in the attic]

Let us say that an element which is interpreted as if it occupied a position other than the one in which we pronounce it has been *displaced* from that position. This usage of displacement, of course, has nothing to do with the very different sense of that term that occurs as one of Hockett's design properties of language (mentioned in Chapter 2). That notion had to do with the possibility of using language to refer to objects and events not present, while the idea at stake here is a purely structural property of sentences.

Displacement is not an unusual property limited to sentences like (21a). Many phrases in sentence types that we produce and understand all the time are systematically displaced from where they are interpreted. What is particularly important is that these displacement relations are not arbitrary, but systematic and deserving of study. It is no exaggeration to say that understanding the possibilities and limits of displacement is central to understanding the grammar of a language, and to the very nature of grammar itself.

Systematic properties of the sort represented by displaced elements are not well described by constituent structures. In terms of phrase structure, an element is either in one position or another, but not both; the question word in (21a) is (pronounced) in a position different from the one it occupies in (21b) for purposes of interpretation.

To reflect the full range of properties of sentence types like content questions (among many others), syntacticians reconstruct the knowledge speakers have of their languages in terms of two structural analyses. One of these, of course, is the analysis of the actual surface sequence of words (the

S-structure of the sentence, where "S" is intended to suggest "surface"). Another is a more abstract representation—more abstract in the sense of being less self-evident from the observable form of the sentence—called D-structure.

The D of D-structure makes sense on the basis of the original name, Deep structure. Many nonlinguists have misunderstood the word *deep* in this expression to refer to something quite different from the role it actually plays in syntax. D-structures, for example, are not themselves the stuff of Universal Grammar, or especially close to a presumed language of thought, or anything else particularly deep. The D-structure of a sentence is an aspect of its technical syntactic analysis within the grammar of a specific language, and it seems wise to avoid the implication that this is somehow more "profound" than other aspects of that analysis, even though the terminology seems somewhat coy and artificial.

D-structure provides a direct representation, among other things, of the ways in which argument phrases are related to particular verbs. It can be viewed as related to the S-structure of the sentence through a series of displacement relations. These are often thought of in terms of the somewhat misleading metaphor of "moving" constituent phrases from one position to another. The distinction between D-structure and S-structure is not intended to imply that the speaker "first" constructs a D-structure, then "moves" something in order to derive an S-structure. The two are simply representations of different aspects of the structure of a sentence: the way in which it is pronounced (S-structure), and the way in which its component phrases are interpreted as arguments (D-structure).

The proper treatment of displacement relations in a formal grammar is a matter of dispute and heated argument among syntacticians. The existence of such relations, however, is not. It is a clear empirical finding on which there is general agreement. Linguistics—syntax, in particular—is a lively and contentious field, and theoreticians have many differences of opinion. The resulting diversity of viewpoints is occasionally appealed to as evidence that all of these issues are simply matters of opinion and aesthetic judgment; but the controversies should not be allowed to obscure the essential agreement that exists on the basic matters under consideration here, regardless of formalism and terminology.

We have no reason to believe that any system of communication in nonhuman animals involves an essential organization of sounds or gestures into something like phrase structure. Still less have we any reason to imagine that abstract structure, related to surface form by patterns of displacement (among other differences) plays any role outside of human natural language. While the specific communicative function of these properties is not immediately apparent, it is worth pointing out that they are are pervasive in human language. In some sense, they are part of the definition of what makes human languages different from other systems, quite independent of whatever explanations we may offer for their presence.

To summarize, then, we apparently need to recognize two distinct structural representations for a given sentence, and a system of rules or principles that relate the one representation to the other. The argument I have just given for this conception of grammar comes from the study of content questions. In these, the apparent lack of one NP (by comparison with the corresponding declaratives) ought to produce massive violation of the argument frame requirements of verbs. This apparent difficulty can be overcome, however, if we say that the argument frame requirement of a verb is something that must be satisfied in D-structure. Content questions have all (and only) the required phrases at D-structure, but one or more of these phrases (in particular, a phrase containing a question word) are displaced from their D-structure position to a different location in S-structure. The sense in which the question phrase is "understood" as filling a particular role is that it occupies that position in D-structure, despite the fact that in S-structure it is found somewhere else.

Many different sorts of facts need to be accounted for in analyzing a sentence. One of these, obviously, is the order in which words appear when we produce it, and the way these words are grouped into phrases. Another critical aspect of sentence structure is the way particular phrases are related to individual lexical items (particularly verbs) that occur in the sentence, so as to satisfy the argument frame requirements of those items. A single phrase structure representation does not provide an adequate representation of both kinds of property at once, so we divide the labor. D-structure is a representation in which the relations between phrases and particular items (especially verbs) of which they are arguments are represented directly. S-structure, in contrast, tells us how the sentence is overtly expressed in English (or some other language, mutatis mutandis). The difference between the two is not arbitrary, but involves precise principles governing relations of displacement.

Facts about the argument frames of verbs are not the only basis for saying that the S-structure of a sentence does not express its entire structure. A different argument for the difference between D-structure and S-structure comes from the way verbs agree with their subjects in number in English.

(23) a. Whose rabbits do you think [e] are(/*is) chasing your cat?

b. Whose cat do you think [e] is(/*are) chasing the rabbits?

The choice of a singular as opposed to a plural form for the verb depends on whether the NP that is its subject is singular or plural. We can only describe the situation this simply, however, on the assumption that the question phrase is considered to occupy the position of subject of the embedded verb (*is/are chasing*). This will be true if we assume that the sentence–initial question phrase is displaced from the subject position indicated by the [e]. We can say that agreement is determined by properties of the phrase that occupies the subject position in D-structure, despite the fact that the phrase is not pronounced in that position.

Finally, a somewhat subtler argument follows from the distribution and interpretation of reflexive pronouns (*himself, herself*, and the like) as opposed to ordinary pronouns. The subscripts in the examples below indicate the intended reference of a pronoun or reflexive. Thus, if two phrases in a sentence have the same subscript i, it is intended to mean that the two refer to the same individual. On the other hand, if one phrase has subscript i and another has $*$ i, this is intended to mean that they *cannot* refer to the same individual. If one phrase has subscript i and another has subscript j, they are intended to refer to two distinct individuals. And so forth.

(24) a. Fred$_i$ changed himself$_{i, *j}$ into a penguin.

b. Fred$_i$ believes Mary$_j$ changed him$_{i, *j, k}$/himself$_{*i, *j, *k}$ into a penguin.

c. Fred$_i$ believes Mary$_j$ changed herself$_{*i, j, *k}$/her$_{*i, *j, k}$ into a penguin.

A reflexive is possible only where it has an *antecedent* (that is, another NP that refers to the same person or thing) in the same clause. Furthermore, simplifying only a little, that antecedent must be the subject of the clause in which it and the reflexive pronoun appear. A nonreflexive pronoun with the same reference is impossible under these conditions and, in fact, reflexive and nonreflexive pronouns are usually in complementary distribution. We might describe these requirements (to a first approximation) as follows.

(25) A reflexive pronoun (and *only* a reflexive pronoun) refers to the same individual as the NP in the subject position in the clause in which it appears.

The relevance of this principle in the present context becomes apparent when we return to our standard example of a construction involving displacement, content questions. Here the NP that acts "as if" it were the subject of a clause for the purposes of (25) may be the question phrase that is pronounced at the beginning of the question, in a sentence like (26).

(26) Which of the women$_i$ does Fred$_j$ think [e] changed herself$_i$/*himself$_j$/ *her$_i$/him$_j$ into a penguin?

Here, the reflexive refers exclusively to the individual designated by the question phrase at the front of the sentence. In the overt (S-)structure of the sentence, this phrase is not in the position of subject of the clause that contains a potential reflexive. In fact, nothing at all is in that position.

This makes perfect sense, in that the question phrase at the front has been displaced from the subject position where it is located in D-structure (indicated once again by [e]). The generalization about reflexive pronouns in (25) can be preserved, but not as a generalization about the surface forms of sentences. It is accurate only in terms of the occupants of positions in D-structure, positions from which they may be displaced in S-structure.

All of these matters (and many others) converge on the conclusion that content questions involve two separate representations. These differ systematically in that a (question) phrase appears in one position in D-structure, but is displaced to another position in S-structure. One representation (D-structure) indicates what is systematic about the relation between a verb and its argument phrases: what constitutes its direct object, its subject, what phrase will determine agreement in number, and so on. The other (S-structure) represents the observable form of sentences: the sequence of words as pronounced. In some cases (many active declarative sentences, for instance), these two are not significantly different, but in others (such as content questions), the two differ systematically.

We have seen that there is no reason to think that any animal communication system displays anything like the difference between D-structure and S-structure. The evidence in favor of that distinction has thus far come from the analysis of content questions, and it might well be objected that, since no animal communication system makes it possible to *form* anything like content questions, such evidence is obviously going to be lacking.

Of course, this is just the point: human language makes it possible to express a vast range of notions that have no analogue in animal communication systems. It is precisely syntactic structure, I maintain, that makes this possible. Without utterances of the relevant kind, evidence for various aspects of syntactic organization will of course be absent. Conversely, it may well be exactly the absence of such structure that limits the expressive capacity of all communication systems relative to that of human language.

Content questions provide a fairly obvious example of differences between D-structure and S-structure, but by no means the only one. Another classic argument for the distinction between these two representations is furnished by the contrast between the two sentences in (27):

(27) a. John is eager to please.
   b. John is easy to please.

On the face of it, the two are entirely parallel. When we look more closely, though, they actually have very different interpretations. In the first, *John* is the one who is going to do the pleasing, and just who he is going to try to please is unspecified. In the second, however, it is *John* who gets pleased, and there is no explicit indication of who does the pleasing. We could propose that the D-structures of the sentences in (27) are something like (28):

(28) a. John is eager [(for John) to please (someone)].
   b. (it) is easy [(for someone) to please John].

In both sentences an unspecified *someone* is assumed in the interpretation but does not correspond to any phrase that is actually pronounced. This situation could be represented as an instance of deletion, another way (in addition to displacement) in which S-structures may differ from their corresponding D-structures. In the *eager* sentence, the D-structure subject of the embedded verb (*please*) is not pronounced, because of its identity with the subject of *eager.* What is immediately relevant, however, is the fact that the S-structure subject of *easy* in the other sentence, the NP *John,* has been displaced from its D-structure position as the direct object of *please.*

Compare these facts with those of a different construction:

(29) John is likely to please the voters of his new district.

Superficially, this sentence is similar to *John is easy to please.* It should be obvious, though, that in (29) the initial NP (occupying the S-structure position of subject of *likely*) corresponds to the subject, not the object of the embedded verb *please.*

The D-structures of the three constructions we have just seen can be contrasted with other sentences having largely similar D-structures but where no displacement or deletion has occurred:

(30) a. John is eager for his new haircut to please his wife.
  b. It is easy (for anyone) to please John.
  c. It is likely that John will please the voters of his new district.

From a comparison of (27) and (28), it should be evident that essentially the same D-structure can be related to more than one S-structure. Conversely, very similar S-structures can be related to quite different D-structures. Consider another classic example:

(31) The chickens are ready to eat.

Sentence (31) has two strikingly different interpretations, but the lack of clarity has nothing to do with ambiguous words or alternative possibilities for organizing words into phrases. One interpretation, that someone is about to eat the chickens, can be represented by saying that the S-structure subject of *are ready* is displaced from the position of object of *eat*. The other interpretation, on which the chickens are the ones who will do the eating, is structurally similar to that of (28a). It involves deletion of the (D-structure) subject of *eat* on the basis of its identity with the subject of the higher verb *are ready*. It is difficult to see how differences like this could be explicated without reference to two distinct kinds of structure for the sentences involved.

The principal point established by these arguments is that the syntax of a language involves more than the mere combination of elements, more indeed than their combination into structured phrases. Sentences involve this kind of structure, to be sure, but they also involve relations of displacement between their overt form (S-structure) and a more abstract representation (D-structure). The latter expresses systematic aspects of a sentence's construction that are important to the way we interpret it but not transparently recoverable from its external form. These are all complications that we, as speakers of a natural language, manage without noticing them. Without them, though, language would not be the flexible instrument of expression and communication that it is—as we can see, perhaps, from the fact that other animals who lack this kind of system can communicate only in much more restricted ways.

The Independence of Syntax

Let us assume that the main points of this description of syntactic structure in natural language are at least roughly correct. What matters is that for any given sentence, we can describe its D-structure and its S-structure, and identify any divergences between the two. Apart from displacement effects, D-structure and S-structure may diverge in other ways, including deletion and insertion. For instance, elements sometimes appear in S-structure that do not correspond to any D-structure item or phrase. Words like *There* in *There's a mouse in the soup!* arguably are inserted in this fashion. The relations between D-structure and S-structure have some remarkable properties, and these have further consequences for our understanding of what knowledge of language involves, and of how human language differs from the communication systems of other organisms.

To see this, let us examine a class of sentences that involve what are known as coordinate structures:

(32) a. Fred bought too many expensive presents and the bank cut off his charge card.
 b. Jennifer sold her old abacus and then bought a computer the very same week.
 c. Pat is majoring in Linguistics and Philosophy.

The word *and* is the key in each of these sentences. It joins two phrases in a way that makes them effectively equal in terms of whatever syntactic function they fill. The phrases involved may be of any sort: these examples involve conjoined sentences, verb phrases, and noun phrases, but other phrasal types are equally possible. A few other conjunctions, such as *or,* play a similar role, but I confine my attention to examples with *and.*

In many cases, much the same meaning as that of a coordinate structure can be conveyed with a different construction:

(33) a. Fred bought too many expensive presents, so that the bank cut off his charge card.
 b. Jennifer sold her old abacus before buying a computer.
 c. Pat is majoring in Linguistics together with Philosophy.

These sentences differ structurally from those of (32). In (33), the second of the two similar phrases is some sort of subordinate structure: a phrase or clause introduced by a "subordinating conjunction" (as our high school English teachers may have called it) or as the object of a preposition.

The difference between the structures in (32) and (33) may seem minor, and there is essentially no change in meaning, but it has important consequences. Assuming that we know what a coordinate structure is, displacements from a position within such a structure are quite narrowly limited. We can formulate the principle that is involved as (34):

(34) *Coordinate Structure Constraint:* A phrase cannot be displaced from a position within one of the conjuncts of a coordinate structure to a position outside that structure.

In forming a content question, for instance, it is not possible to displace the question phrase from only one of the conjoined phrases shown in (32):

(35) a. *What did Fred buy [e] and the bank cut off his charge card?
     b. *What did Jennifer sell her old abacus and then buy [e] the very same week?
     c. *What is Pat majoring in [e] and Philosophy?

When we consider the semantically similar sentences in (33) that do not involve genuine coordinate structures, we see that displacement from only one part of the structure is perfectly acceptable:

(36) a. What did Fred buy [e] so that the bank cut off his charge card?
     b. What did Jennifer sell [e] before buying a computer?
     c. What is Pat majoring in [e] together with Philosophy?

It seems, therefore, that something quite specific about conjoined structures prevents displacement of a phrase out of them. Since displacement is not blocked out of other structures with essentially the same meaning, the problem cannot relate to their meaning. Apparently, it must be an aspect of their syntactic form.

When we ask ourselves why this should be the case, the situation is puzzling. Think of language as a system that relates sound and meaning to each other. If that is all that is involved, then a sentence that the rules of the grammar would appear to produce, but that is still unacceptable, ought to be excluded because either its phonology (sound) or its semantics (meaning) is somehow defective. In these cases, though, neither explanation seems to work. It is always possible that we have overlooked something else and that there might be an explanation for the ungrammaticality of the sentences in (35) in terms of their surface form, but this line of reasoning does not look very promising.

There certainly seems to be no problem with the meaning of the prohibited sentences in (35); it is possible to express the intended meaning of each in another way, showing that whatever the problem may be, it is not that the meanings are ill formed:

(37) a. Fred bought something, and the bank cut off his charge card—what was it?
   b. Jennifer sold her old abacus and then bought something else the very same week—what was it?
   c. Pat is majoring in something and Philosophy—what is it?

Or we can just leave the question word in place, to produce an "echo question":

(38) a. Fred bought WHAT and the bank cut off his charge card?
   b. Jennifer sold WHAT and bought a computer?
   c. Pat is majoring in WHAT and Philosophy?

What seems to be going on here is a restriction on well-formed structures that is *purely* a matter of syntax: the constraint in (34) says that any sentence with a specific pattern of displacement is excluded, and this has nothing to do with either sound or meaning. Just syntax.

We might be tempted to think that the problem in (35) has to do with the formation of content questions, but in fact the limitation expressed by (34) applies equally well to *all* constructions that involve displacement. A number of these, some of which we have seen before and some of which we have not, are illustrated in (39). I will not analyze each of them, but it should be apparent from what we have already seen that something is displaced in each case; the D-structure positions of the displaced phrases are shown as usual with an [*e*]. For each construction, we can contrast a sentence involving a violation of the constraint with a very similar structure that does not violate (34).

(39) a. i. This is the kind of chips that I like [*e*] with guacamole.
       ii. *This is the kind of chips that I like [*e*] and guacamole.
    b. i. This class is hard to keep up with [*e*] at the same time as Economics.
       ii. *This class is hard to keep up with [*e*] and Economics.

c. i. Bagels, I like [e]—but smoked salmon, I've never developed a taste for [e].

   ii. *Bagels, I like [e] and smoked salmon.

d. i. Big and tough as he is [e], Fred still makes a poor linebacker.

   ii. *Tough as he is big and [e], Fred still makes a poor linebacker.

e. i. The thunder seems [e] to be coming this way along with the lightning.

   ii. *The thunder seems [e] and the lightning to be coming this way.

f. i. The subject and the object cannot be displaced [e] at the same time.

   ii. *The subject cannot be displaced [e] and the object at the same time.

Constraint (34) applies regardless of the kind of argument a displaced phrase represents. Some examples in which phrases are displaced from various grammatical positions in the process of forming relative clauses are given in (40). In each case, the sentence is ruled out if the displaced element comes from one of the conjoined phrases in a coordinate structure. Prescriptive grammarians object to "stranded" prepositions like the *to* in the indirect object examples, but the construction is a fact of life in modern American English. Even prescriptivists should find a distinct difference of grammaticality between the two sentences of the pair, which is the point of the example:

(40) *Subject:*

   a. The man who [e] sold me a new Volkswagen for $2,000 was bald.

   b. *The man who your friend and [e] sold me a new Volkswagen for $2,000 was bald.

Direct Object:

   a. The car which the bald man sold me [e] for $2,000 was a lemon.

   b. *The car which the bald man sold me a new truck and [e] for $2,000 was a lemon.

Indirect Object:

   a. The person who the bald man sold the Volkswagen to [e] for $2,000 was a sucker.

   b. *The person who the bald man sold the Volkswagen to [e] and his mother for $2,000 was a sucker.

Oblique:

a. The price which the bald man sold the lemon to the sucker for [*e*] was $2,000.

b. *The price which the bald man sold the lemon to the sucker for [*e*] and a long series of low monthly payments was $2,000.

Furthermore, the D-structure position from which the relative pronoun has been displaced can be a long way from where it appears in S-structure. In fact, there is no limit (in principle) to how deeply embedded that position can be.

(41) a. A person who I believe that Fred told me that Mary was under the impression that she had read in the paper that you would be able to identify [*e*] is wanted for questioning about a lost dog.

b. A person who I believe that Fred told me that Mary was under the impression that she had read in the paper that you would be able to save yourself trouble {by identifying [*e*]/*and identify[*e*]} is wanted for questioning about a lost dog.

Clearly we are dealing with a limitation on well-formed displacements *in general,* not with a set of properties of the constructions used to form relative clauses, content questions, and so on. And that is a surprising fact, because displacement is purely a property of syntax. It is a fact about the relationship between D-structure and S-structure, and not a fact about either meaning or surface form in itself. That such a condition can play a role in deciding whether a sentence is acceptable or not, quite independent of its meaning or surface form, shows that knowledge of syntax has a place of its own in the overall fabric of language.

The principle in (34) is not the only one that limits the possibilities for displacing phrases. Another such constraint blocks displacement from a position within a sentence that is itself part of an NP containing a determiner and a meaningful head noun. Its effects can be seen in (42). In each example to follow, the NP from which displacement has taken place is shown:

(42) a. Which new word-processing program did Kim write the manual for [*e*]?

b. *Which new word processing program did Kim write [NP the manual that documents [*e*]]?

As in the case of violations of (34), this limitation on displacement applies to a variety of constructions:

(43) a. *Fred is not the only man that/who(m) I know [NP a woman who is in love with [e]].

   b. *Syntax is hard to read [NP the book Lightfoot and I wrote about [e]].

   c. *Smoked bluefish I can show you [NP a store that refuses to sell [e]] any more.

   d. *Tall as he has [NP a younger bother who is [e]], Fred still can't make his free throws.

We can also easily confirm that this constraint applies to displacements from a variety of positions.

(44) *Subject:*

   *Which manual did [NP the discovery that Kim wrote [e]] convince you to hire her?

Indirect Object:

   *Which bicycle did you try to sell your old car to [NP the man who was riding [e]]?

Oblique:

   *Which policeman did you want to see a man about [NP a dog who bit [e]]?

And finally, the prohibited displacement site can again be from a D-structure position arbitrarily far from the S-structure position of the displaced phrase:

(45) Who do you believe that Fred told Mary that he was under the impression that he had read {a description of [e] in the paper/*[NP a newspaper article that described [e]]}?

Syntactic research since the late 1960s has uncovered a number of principled constraints on syntactic displacement, of which the two just illustrated are among the best known. Common to all of them is the fact that displacement is restricted for reasons internal to the syntactic system, under circumstances that can be formulated precisely in terms of syntactic structure but that have no coherence in terms of sound or meaning.

Perhaps conditions like (34) and others will turn out to be separate principles of grammar, or perhaps we will eventually find a unitary condition that includes all of them. In the present context, that does not matter. What does matter is the fact, that as far as we can tell, whatever the constraints may ultimately be, they will still be purely syntactic in nature. They are not, apparently, grounded in sound or meaning—and they have no analogues in any system other than that of human natural language.

With regard to restrictions like (34)—although it would take us too far afield to demonstrate it here—constraints on displacement of the sort we have been exploring are not just facts about English. Several decades of investigation support the claim that they are true of *all* human languages (or at least of all the substantial number that have been examined thus far). This kind of knowledge could not plausibly have been acquired on the basis of experience (try to remember an occasion on which your parents taught you to obey the coordinate structure constraint, for example). Therefore it seems likely that these aspects of syntactic organization are as much a part of the biologically determined human language faculty as the structure of vervet monkey alarm calls is specific to animals of that species.

If we believe there is nothing very special about human language, that all of its properties follow from its being a system for relating sound and meaning, then we would expect everything about it to derive from aspects of sound and/or meaning. And of course a great deal does, once we get the system worked out. Still, a great deal does *not,* and this much at least remains an undigested residue of purely linguistic facts—facts that as part of our knowledge demonstrate that this knowledge has an interesting and autonomous structure.

We see that syntactic organization is a great deal more than a matter of putting together meaningful symbols (words, signs, whatever). It involves intricately organized structure, and principles that affect the expression and interpretation of meaning in ways that derive from that structure itself. These structures and principles are not, of course, immediately apparent, even though we unthinkingly employ the structures and obey the principles. They follow from our very nature as human beings, and from the language faculty that we deploy as a result.

Much of what we have uncovered about syntactic structure may seem incidental to the fundamental role grammar plays in language: allowing us to achieve an unlimited range of expression on the basis of a limited set of symbolic items. There is surely an irreducible minimum of structure, without which this expressive flexibility could not be attained. For example, without the property of recursion, a communicative system could not come near to filling the role language plays for us.

Perhaps an unbounded range of complex meanings could be expressed without displacement relations, without constraints on the scope of displacement, or in the absence of various other specific properties of human language. We really cannot tell, because the only systems we know that display that capacity, namely human natural languages, all have those properties. These products of the human language faculty have much in common that is striking and specific. Perhaps they are some sort of accident of evolution; but perhaps the properties we find are necessary concomitants of a computational system that relates sound (or signs) and meaning in the general way human languages do.

Something about the cognitive organization of human beings allows us to make use of a communication system that involves the kind of syntactic organization we have been discussing. If we wanted to show that another species was capable in principle of mastering the essentials of human language, a question we address in Chapter 10, it stands to reason that we would have to demonstrate that those animals were able to acquire and use these same structural properties. Meaningful symbols, even combined with one another in limited ways, are not all there is to human language. They are not even the most essential part, when it comes to the unbounded flexibility of that instrument.

9

*Language Is
Not Just Speech*

"All animals have some kind of a language. Some sorts talk more than others; some only speak in sign-language, like deaf-and-dumb. But the Doctor, he understands them all—birds as well as animals. We keep it a secret, though, him and me, because folks only laugh at you when you speak of it."

—The Cat's-Meat Man, from *The Voyages of Doctor Dolittle*

Apart from the honeybee dances of Chapter 4, most of the communication systems we have considered up to this point share Hockett's design feature of a vocal-auditory channel. Frogs croak, birds sing, monkeys chutter to one another. All of these (as well as the long-distance subsonic rumbling of elephants, the songs of whales, and many others) involve one animal's producing sound from which another animal derives information. Sound seems to be a peculiarly natural medium in which to communicate.

When a male stickleback provides a visual signal to females of his interest in mating by changing the color of his belly, the latency of the message (in terms of the time it takes to turn it on and off), as well as the range of messages conveyed, obviously provides a vastly less flexible channel of communication than that of sound. Some species of squid are renowned for the complex patterns of color they can produce on their bodies, as well as the rapidity with which these patterns change; but even if we had some notion of just what messages these patterns might be conveying, it is clear that the specialized apparatus involved makes this particular use of a visual channel the exception rather than the rule.

Yet it is certainly not the case that sound has a monopoly on communication. Even relatively simple life forms like bacteria receive (and convey) information via chemical senses similar to smell and taste. Every species with eyes uses them to derive information about the surrounding world, and practically every such species also engages in behavior that provides information to conspecifics through their ability to see one another. One monkey adopts a particular gait in approaching a goal; other monkeys observe it and gauge the first individual's degree of purposiveness and aggressiveness in deciding whether or not to challenge him.

One could, of course, regard this sort of activity as lacking in "communicative intention." The first monkey might indeed intend to intimidate his competitors by his show of determination, but it is at least as likely that his action is merely the natural outward reflection of his internal state. Visually based communication in familiar species is by no means limited to this case, however, as we have already seen.

The natural communicative behavior of primates (especially the great apes) involves a number of expressive gestures that unambiguously express the intention of one individual to communicate something to another. Since primates have arms and legs that are quite freely movable and not constrained to serve exclusively the demands of locomotion, hunting, or eating, much of this gestural expressiveness involves limb movements, especially of the arms and hands.

Being primates ourselves, we are hardly lacking in comparable abilities, even if we also have natural language to serve our communicative needs. We wave goodbye, thrust our fist in the air in triumph, show our hand to another with middle finger extended to taunt or enrage, give a thumbs-up signal, and much more.

Taken as communication systems, however, all of these gestural activities on the part of primates are structurally no more intricate than a set of bird calls. They are simply an inventory of some relatively limited, finite number of discrete messages, each conveyed by a specific gesture distinct from the other possibilities. They may be modulated in intensity to correspond to the precise degree of the internal state that gives rise to them in the first place, and they may be used appropriately in a variety of circumstances, but they still do not constitute an open-ended, flexible system in terms of which we can construct genuinely new messages.

When we turn from these natural gestures, each used relatively generally by the members of a species, to the special case of the sign languages used by hearing-impaired humans (and some of their hearing friends, colleagues, and relatives) to circumvent a perceptual limitation, no truly fundamental difference strikes us at first glance.

A note on terminology: As has become conventional, the word *Deaf*—with a capital D—is used here to refer to individuals who share a distinctive culture, and whose principal means of expression is a signed language such as ASL. Some individuals who participate in the Deaf community may have normal hearing. They may, for instance, have learned to sign from Deaf parents, in which case they may well have fully native control over their signed language, as well as one or more spoken ones. With a lowercase d, the word *deaf* simply refers to an individual with a severe hearing impairment.

It is natural to assume that those who cannot hear what others are saying in a language like English must turn of necessity to a greater reliance on the natural gestures available to us all, and a signing person's behavior may seem to be just that. On this understanding of their nature, signed languages would be nothing more than an elaboration of natural gestures, iconic evocations of the surrounding world. Even though such a system might well be extended far beyond the ordinary, with a substantially wider variety of conventionally interpreted gestures than usual, it would still have the fixed, finite character of other gestural inventories. Formally, it would differ from a set of bird calls only in size (and of course medium of expression), not in structural complexity.

Signed languages were generally viewed in exactly this way prior to 1960, an extremely recent date in the overall history of humankind's understanding of language and communication. At that time William Stokoe began to publish the results of his linguistic analyses of American Sign Language. Stokoe's work revolutionized our understanding not only of ASL but also of the human language faculty more generally. It opened up the possibility that the properties and complexities of languages such as English, Japanese, and Navajo were not restricted to spoken languages, but might be found in another modality as well. Exploring that insight over the subsequent decades has provided resounding confirmation of the correctness of this conclusion.

Signed Languages and Their Study

Most of the research devoted to the structure of signed languages has focused on American Sign Language, but ASL is only one instance of a language that operates in a manual-visual, rather than an auditory-aural, modality. Signed languages are at least potentially as distinct from one another as spoken languages are, in clear contrast to natural gesture systems, which are largely the same across the species.

For obvious reasons, communities of Deaf individuals do not display the kind of continuity over thousands and thousands of years that hearing communities do. As a result, there are no signed languages that have been transmitted over many generations, with the consequent opportunity for the natural processes of language change to have their usual cumulative effects. Signed languages are in general of relatively recent origin, and since they tend to develop much of their lexical stock from iconic bases, accidental resemblances between signs with similar meanings in different languages are rather more common than is the case in spoken languages. Given a chance, though, signed languages develop all the same arbitrariness in the relation between form and meaning that we find in spoken languages, and two communities sharing an original common language diverge in essentially the same ways regardless of whether that language is signed or spoken.

At first glance, the very notion of signed languages may sound somewhat contradictory: languages are spoken and heard, after all, are they not? If we do not have (signing) Deaf relatives or friends, we are likely to think of "sign language" in one of two ways. On the one hand, it might be a rather rudimentary inventory of gestures that either point to or mimic objects and events in the world, gestures that can be strung out one after another but that lack any real structure. On the other hand, it might be English (or some other spoken language), with some hand waving substituted for, and basically representing, each of the words. Neither of these conceptions is remotely close to the truth.

Before proceeding much further, we should eliminate another, less central misconception. We usually speak of ASL and others as "manual" languages, in the sense that they are spoken with the hands. This is a significant oversimplification. In actuality, many movements of parts of the body other than the hands are important in signing. Researchers have understood this since the mid-1970s and have uncovered a rich variety of grammatical roles for facial expressions produced simultaneously with manual signs. Portions of these expressions look strictly iconic, but many aspects have been incorporated into the grammar.

For example, a particular configuration of the eyebrows marks relative clauses; another distinctive facial configuration accompanies (information) question words such as WHO? WHAT? WHERE? Yes or no questions (*Are you a policeman?*) differ from the corresponding assertions (*You are a policeman*) solely in terms of facial expression. The role of facial expression in signing is comparable to the role of intonation in spoken language: some of it is grammaticalized and discrete, while other aspects are continuous and emotive.

The signer also shifts facial orientation to reflect differences in point of view. An example is the perspective reflected in reported speech (when I say *Fred said that he was a policeman*) as opposed to direct speech (Fred said *I am a policeman*). The axis of gaze between signer and addressee is essential to anchoring the system of demonstratives and pronouns. All of these matters have great importance in the way such languages work, although most attention has focused on the activity of the hands.

William Stokoe, whose work initiated the linguistic analysis of signed languages, was trained as a linguist. As often happens, he did not find work as a linguist, but instead was hired to teach English literature at Gallaudet University in Washington, D.C. Working with the Deaf, he became convinced that they indeed had a system that was amenable to study by the methods applicable to other languages. Stokoe developed an analysis of the expression system (which we will call the phonology) of ASL.

It quickly became apparent to Stokoe and later researchers that signed languages have a rich and elaborate structure of their own. They have not only semantics (as we might expect), but also syntax, morphology, and phonology. The last might better be called something else perhaps, because the *phon-* of *phonology* means "sound," but none of the other terms people have suggested (such as Stokoe's "cherematics") have caught on.

Further, the abstract organizing principles of the phonologies of signed languages have a great deal in common with those of the phonologies of spoken languages — rather more than we might think. For instance, the current research literature contains much discussion of the role of the syllable in ASL. The kinds of representations that are significiant in theories of the phonologies of spoken languages have very similar functions in the analysis of signed languages. As a result, there seem to be valid reasons to maintain the terminology in spite of the apparent paradox.

A Little History

Signed languages are completely natural linguistic systems that (can) arise naturally within Deaf communities. The history of ASL is actually a bit more complex, though. Its origins are in a naturally originated signed language used in France that was adopted (and adapted, codified with indifferent success) by French teachers of the deaf.

With the emergence of the modern city in Europe came the natural rise of communities of deaf people in Paris, Madrid, and elsewhere. In the eighteenth century, the Abbé de l'Epée learned a signed language that he used to teach French to the deaf in France. The system was further developed for deaf education there.

In the early nineteenth century Rev. Thomas Gallaudet went to Europe in search of methods to apply in the education of a deaf child of his acquaintance. While attitudes in England were generally oralist (oriented toward the teaching of spoken rather than signed language), he found a more sympathetic climate in France. He persuaded the French Deaf instructor Laurent Clerc, from the Institute of Deaf-Mutes in Paris, to return with him to the United States. There, in 1817, they set up the American Institute for the Deaf (originally called the Connecticut Asylum for the Education and Instruction of Deaf and Dumb Persons) in Hartford, Connecticut.

The introduction of the French signed language in Hartford led to the opening of many more schools for the deaf in the United States based on the same approach. At the time, signed languages had emerged in individual communities, without any of them becoming a de facto standard. One was widely used on Martha's Vineyard, which had a high incidence of congenital deafness at least through the first half of the twentieth century. The Deaf formed so large and important a part of the community, in fact, that many hearing people on the island also learned to sign, sometimes doing so among themselves in preference to speaking English.

The French sign system quickly took over, displacing (and to some extent incorporating) indigenous signing systems and eventually giving rise to American Sign Language. ASL has since become the standard language of the North American Deaf community, although distinctive dialects have developed in relatively isolated subcommunities.

In 1864 Congress authorized a national college for "deaf-mutes." This institution subsequently became Gallaudet University, named after Thomas Gallaudet and his son Edward (the first principal of the new college). It is still the only liberal arts college for deaf students in the world, though there are also technical schools, such as the National Technical Institute for the Deaf at the Rochester Institute of Technology.

In the 1870s much of the progress based on the introduction of ASL in the Deaf community was arrested by the rise of oralism, which in general actively discouraged signing. Alexander Graham Bell played a principal role in this movement. Apparently both Bell's mother and his wife were deaf, although neither ever acknowledged the fact. He himself became deeply involved in the issue, and his scientific prestige contributed to the effective elimination of signing from the curriculum of institutions for the deaf following a major conference of educators in Milan in 1880. The oralist movement is still with us, and it is only since the 1960s that the use of sign and the role of deaf teachers in schools for the Deaf has returned to anything like earlier levels.

Signed languages are structurally quite independent of any substantial basis in the spoken language of the surrounding hearing community. For instance, standard sign languages exist in the United States (ASL), Great Britain (BSL), Australia (Auslan), and New Zealand (NZSL), all regions where English is widely spoken. An American traveler in any of these countries, or an Antipodean in New York, will encounter some words that may seem eccentric, but by and large will have no trouble understanding the natives—or being understood.

British, Australian, and New Zealand sign are mutually intelligible in the main, though there are distinct differences. On a standard list of one hundred basic lexical items, any two of these languages will share about eighty or more. This commonality results from the basis that both Auslan and NZSL have in BSL. Deaf immigrants introduced BSL to Australia and New Zealand in the nineteenth century, though the language evolved in slightly different ways in the two countries.

ASL, in contrast, is not mutually intelligible with any of these languages. Of the same list of one hundred basic vocabulary items, ASL shares at most thirty or so with any of the others. On the other hand, ASL (but not BSL, Auslan, or NZSL) is mutually intelligible with French sign language, at least to some extent. Together with Irish Sign Language and LSQ (Langue des Signes québecoise) in Quebec (as a result of the influence of the Roman Catholic church), these have a common origin in the eighteenth- and nineteenth-century French sign mentioned above. In another example, Taiwanese Sign Language is largely mutually intelligible with Japanese Sign Language, but not (apparently) with the sign languages of the rest of China or Hong Kong.

The Question of Iconicity

A common (mis)impression about signed languages is that signs, unlike words, are direct, quasipictorial representations of the things they symbolize. That is, signs are commonly thought to be *iconic* in representing their referents directly, rather than arbitrarily.

It is true that iconicity plays a larger role in signed languages than in spoken ones. New signs are often created on an iconic basis, and since most sign languages are of relatively recent origin, the signs have not had as much chance to erode as the words of spoken languages. In any case, the difference between signed and spoken languages with respect to iconicity is not at all absolute. Spoken languages have a certain amount of iconicity: think of words like *ping-pong* and at least some onomatopoetic words such as *bang, thud, whack*. And then there are the sounds animals make. If you doubt that these are actually words in a language, ask yourself why English roosters salute the dawn with *Cock-a-doodle-doo!* while their French counterparts say *Cocorico!*

We see immediately that signs cannot be completely iconic and must have an arbitrary aspect, since the signs for a given concept may be quite different in different languages. As an example, consider the signs for FATHER in ASL and C(hinese)SL in Figure 9.1.

Conversely, the same sign may indicate quite different things in different signed languages. The ASL sign CHURCH is more or less the same as the Mexican Sign Language EGG. In *The Signs of Language,* Edward Klima

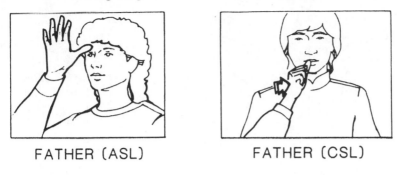

FATHER (ASL)   FATHER (CSL)

Figure **9.1** FATHER in ASL and Chinese Sign Language

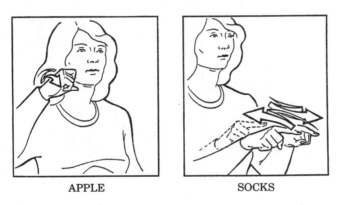

APPLE   SOCKS

Figure **9.2** Two signs that are not transparently iconic

and Ursula Bellugi provide a detailed comparison of many signs in ASL and Chinese Sign Language. They show that some signs are identical in the two languages but have different meanings. Some signs in one language are possible but nonoccurring signs in the other. Still others occur in only one of the two languages and would be impossible in the other, because they violate language-specific constraints on sign formation.

Signed languages have a considerable number of patently arbitrary signs. That simply has to be the case for a sign like FAKE, FALSE, whose meaning is abstract and nonpictorial. It can even be true for fairly concrete signs, such as those in Figure 9.2.

Even where we can identify iconic motivation for a sign—tell a sort of creation myth about it, as it were—it is still a conventional symbol, not just a picture. Compare the ASL, Brazilian SL and (New Delhi) Indian SL signs for WHITE in Figure 9.3. The first two are based on a reference

ASL   BSL   ISL

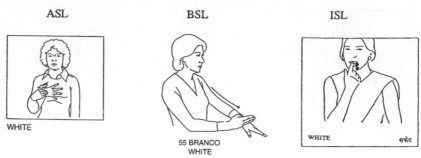

Figure **9.3** WHITE in ASL, B(razilian) SL, and I(ndian) SL

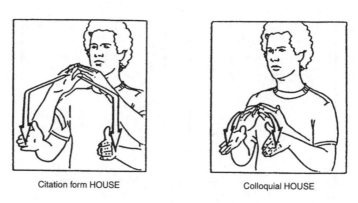

Figure **9.4** Suppression of iconicity in formal and colloquial ASL HOUSE

to white shirts, a fact that is probably not obvious from looking at them. In any case, the two are quite different and not mutually substitutable. And the Indian sign is a reference to white teeth. All three are iconic in origin, but all three have the meanings they do by virtue of their status as conventionalized parts of individual languages.

In many cases we may be able (with a little prompting) to identify the iconic basis of a sign when it is produced in isolation. Signs in fluent conversation, however, may involve substantial suppression of this aspect of their structure, as movements are simplified in the flow of signing. The formal and colloquial forms of ASL HOUSE in Figure 9.4 provide a simple example.

In studies conducted in a number of communities, researchers have asked people to guess at the meanings of signs. Hearing subjects who know no signed language can usually guess the meanings of a small number of signs in such a list at least some of the time, but most signs are impossible

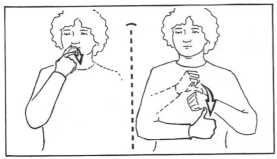

RED~SLICE[+] meaning 'tomato' in 1918-ASL   TOMATO in M-ASL

Figure **9.5** (1918 ASL) RED + SLICE → (Modern ASL) TOMATO

to surmise (and the signs used in these tests are generally simple concrete nouns and actions, where the relation between the sign and its referent is most likely to be apparent). Even when the task is made multiple choice, performance hovers at the level of chance. Interestingly, when the subjects are themselves signers (but signers whose language is unrelated to that from which the stimuli are drawn), they do a bit better but still come nowhere near to generally correct guessing. And on tests of recall, signers do not recall iconic signs significantly more frequently than noniconic ones.

Historical change typically reduces or eliminates the iconic basis of signs, from which we conclude that such a basis cannot be essential to the sign's meaning. Often an original compound is simplified to a single sign that involves elements taken from each part of the original compound. The result typically shows no respect at all for whatever iconicity may have been present earlier. For instance, in ASL as of about 1918, the sign for TOMATO was a compound RED-SLICE. In contemporary ASL, the compound has been reduced to the single sign on the right in Figure 9.5.

The kinds of error that children make when acquiring signs show that the iconic basis is not crucial. For example, the ASL sign FLOWER is made with a movement of the hand from one side of the nose to the other, and plausibly derives from an iconic representation of smelling a flower. That basis was evidently not apparent (or not relevant) to a child described by Richard Meier who consistently mispronounced this sign with the same gesture located at the ear instead of the nose.

Appropriately enough, then, signs exemplify Ferdinand de Saussure's principle of *l'arbitraire du signe* just as the words of spoken languages do, regardless of any historical basis (for either) in iconic representations of the world. With that background, we can now explore the internal structure of a signed language, and the consequences of a visual as opposed to an auditory modality for language more generally.

Components of a Signed Language

Like any other language, spoken or signed, ASL displays systematic grammatical organization of a number of distinct sorts. Parallel to the phonology of a spoken language, ASL has a set of basic formational elements, meaningless in themselves, which are organized into individual signs. Regularities specific to an individual language govern how this combination takes place, which combinations are valid and which are not, and other structural properties. Just as words in a spoken language can combine several meaningful elements (for example, [[*mean+ing*]+*ful*]) according to the principles of the language's morphology, complex signs can contain multiple parts each with its own meaning. The principles of syntax combine signs (whether morphologically simple or complex) into larger phrases and sentences. In each of these areas we find a combination of universal constraints and principles with language-particular choices. The principles we observe are generally shared with spoken languages, to the extent that they are not based directly on the properties of sound or vision. These commonalities place signed languages squarely within the overall range of possibilities allowed by the human language faculty.

Phonology

To a nonsigning observer naive about the language, individual signs appear to be unitary gestures—unlike spoken words, which are made up of combinations of sound segments. The power of the phonological system of a spoken language resides in the ability of a relatively small number of basic elements (the sound types, or phonemes) to combine and recombine in different ways to make a vast range of distinct words.

Phonemes can themselves be regarded as combinations of contrasting values along a small number of dimensions. To take English as an example, we have sounds that differ in terms of *place of articulation.* Some are formed at the lips ([f, v, m, p, b]), others by the tongue along the teeth (the two *th* sounds of *bath* and *bathe*) or alveolar ridge ([s, z, n, t, d]), and still others by the tongue body at the velum ([ng, k, g]). Crosscutting this difference is that between voiced sounds (those involving vibration of the vocal folds, such as [v, b, z, d, g]) and voiceless ones ([f, p, s, t, k]). Another dimension is that on which nasals (sounds with the velum lowered, such as [m, n, ng]) are distinct from corresponding nonnasals ([b, d, g]). With several others, a single set of such parameters of sound can be constructed that serves for all of the world's languages (though of course not all languages make the same use of all of its aspects). As in syntax, the genius of the phonological system is its exploitation of a discrete combinatorial system to make a large number of ultimately distinctive linguistic elements out of a limited set of basic resources.

We also describe ASL signs as made up of specific selected values on each of a few basic formational parameters. Stokoe's original analyses in the 1960s identified some of these, though not necessarily in precisely the form today's investigators would prefer. The basic parameters of sign formation correspond to differences in *handshape*, *place* of articulation, and *movement*. Examples of signs differing along only one of these dimensions at a time are given in Figure 9.6.

Several other parameters of signs have been identified. Klima and Bellugi's book presents differences in *orientation*, *plane* (of movement), *contact region*, *hand arrangement* (one hand versus two), among others. Without repeating their discussion, let us say that the set of dimensions along which signs differ is entirely comparable to the set of component properties that distinguish sounds in spoken languages.

Any particular spoken language utilizes only a subset of the values for a given parameter that are found in the full range of the world's languages. For instance, with respect to the position of the larynx and its effects relative to a consonant, English basically shows only a difference between voiced sounds (like *b, v*) and voiceless ones (like *p, f*). When we look closely, we see that the voiceless stop consonants [p, t, k] show an additional possibility. When—and only when—they occur at the beginning of a stressed syllable, they are *aspirated*, or followed by a short puff of air. This action results from a distinct kind of laryngeal behavior, and some languages treat the difference between aspirated and unaspirated consonants as being just as distinctive as, say, voicing in English. Many other languages have consonants that are formed with a simultaneous constriction of the vocal folds, similar to the glottal stop in the middle of English "Uh-oh!" Glottalized consonants are one of the possibilities employed distinctively in human languages in general, though not in English.

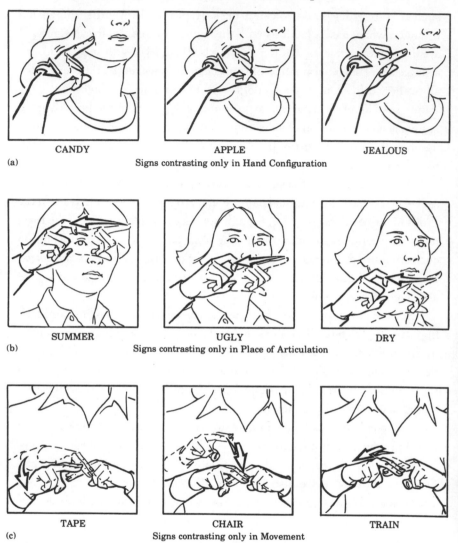

CANDY   APPLE   JEALOUS

(a)   Signs contrasting only in Hand Configuration

SUMMER   UGLY   DRY

(b)   Signs contrasting only in Place of Articulation

TAPE   CHAIR   TRAIN

(c)   Signs contrasting only in Movement

Figure **9.6** Minimal contrasts illustrating major
formational parameters of signs

Similarly, not all the possible values for sign parameters may be allowed in any particular signed language. For example, only certain handshapes are possible in ASL signs. Figure 9.7 shows two very similar possibilities, one of which is used in ASL and the other in C(hinese) SL.

A single sign may involve one or both of the hands (as in the contrast

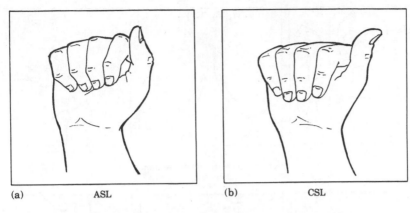

(a) ASL (b) CSL

Figure **9.7** The closed-fist handshape in ASL and CSL.
Note the differences in thumb placement and hand closure.

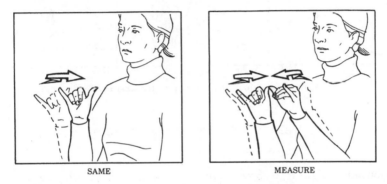

SAME MEASURE

Figure **9.8** Contrasting signs that differ only in the number of hands used

illustrated in Figure 9.8). If both hands are involved in a sign, though, one
(the "weak" hand) must either be doing the same thing as the "dominant"
hand, or else it must remain stationary. These two possibilities are illustrated in Figure 9.9.

We have said nothing about other formational elements of signs, including the range of permitted movements, or about limitations on the positions in space within which signs are articulated. All of these aspects of
the "phonetics" and phonology of signed languages (and many more) have
been the subject of a lively and substantive research literature in the years
since Stokoe's original discoveries.

These formational elements combine with one another as separable
components of the sign. Saying this raises the question of how we know that

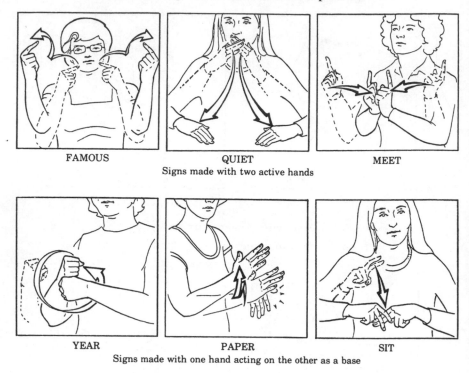

FAMOUS QUIET MEET

Signs made with two active hands

YEAR PAPER SIT

Signs made with one hand acting on the other as a base

Figure **9.9** Possible two-handed signs

the division of signs into components has any linguistic reality, rather than being an invention of the linguist. The same issue is equally salient in the analysis of spoken languages: how do we know that the component sounds of spoken words play a role in the speaker's organization of language, not just the linguist's?

Insofar as we write in an alphabetic system, we have been brought up to believe that words are composed of a sequence of sound units. That is not the only possibility, however. In Chinese orthography the structural unit represented by a single character is more or less the same as a syllable, not a phoneme. Korean Hangul orthography, on the other hand, appears to organize the phonemes of the language into a set of individual features of contrast, such as the difference between voiced, voiceless, and glottalized sounds.

Still, we do have evidence for the correctness of the "alphabetic" analysis of words. Consider the character of speech errors. Slips of the tongue suggest that phonetic segments serve as units in the planning of utterances.

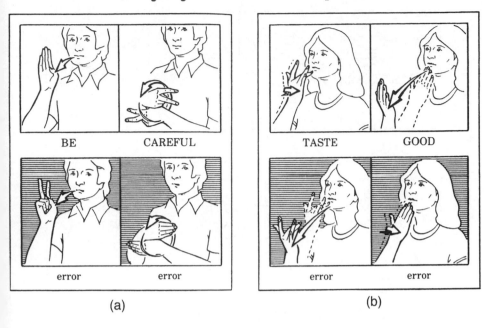

BE CAREFUL   TASTE GOOD

error error   error error

(a)   (b)

Figure **9.10** Signing errors involving transposition of (a) handshape and (b) movement. The erroneous sign is shown below the correct sign.

Thus, a speaker who intended to say "left hemisphere" said "heft lemisphere" instead, transposing the /l/ and the /h/. In general, the units that transpose in slips of the tongue correspond to the units of phonological description: features, phonemes, syllables. Since elements of roughly the granularity of letters of an alphabetic system play such a part in the planning of speech (even in languages with no writing system, or one based on a different principle), we infer that they are real parts of the organization of language.

The same kind of evidence is available for signed language. Like speakers of English, ASL signers sometimes make "speech errors," and when these occur they typically involve transpositions of the basic formational elements that make up signs. Figure 9.10 illustrates signers' errors involving exchange of (a) handshape and (b) movement.

We also find language-specific restrictions on the ways in which elements can combine with one another, and the fact that combinations of features or segments are referred to in such rules furnishes additional evidence for the significance of these units. Just as English allows initial *st* and *sp*

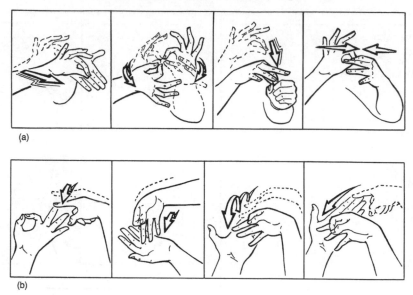

Figure **9.11** Signs using the pinching hand configuration in
(a) ASL and (b) CSL

but Spanish does not (*stamp* versus *estampilla*, *Spain* versus *España*), signed
languages can differ in the combinations they allow.

Both ASL and CSL use a "pinching" hand configuration, in which the
thumb and index finger are in contact and the other three fingers are extended. In ASL, when the hand with this shape makes contact with the
other hand, the contact region is always the touching tips of the thumb and
index finger, as in Figure 9.11(a). In CSL, however, as shown in Figure
9.11(b), the part of the pinching hand that makes contact is one (or more) of
the *other* three fingers, rather than the "pinch" itself. Since these constraints
differ from language to language, what counts as a well-formed sign in one
language may correspond to a different sign in another language, or perhaps to a possible but nonoccurring sign, or to something that could not be
a sign at all in that language.

The same phonetic shape can be identified with quite different words
in two spoken languages. For instance, the Georgian word [kʰɪṭi] "kind of
wooden spoon" sounds virtually the same as the English *kitty*. The Georgian word contains a consonant type (the glottalized *t* in the middle) that
does not occur in English, but a naive English listener hearing this word in
isolation would probably interpret it as *kitty* pronounced a little oddly.

As we study signing in more detail, many properties emerge to indicate that the phonology of ASL is just as rich, and just as much a part of the identity of the language, as the phonology of English or French or Georgian. When the phonological constraints are violated, the result may range from complete unintelligibility to differences that give the signer a "foreign accent." (Such would be the case, for instance, if a signer substituted the CSL version of the closed-fist handshape in Figure 9.7 for the ASL version.) The phonology of a language is very much a part of its basic character.

Morphology

In the domain of word structure, or morphology, there is also more to sign language than meets the untutored eye. Consider the ASL signs for pronouns. "I" is signed by pointing to oneself with the index finger; "you" by pointing at the hearer; and "he/she/it" by pointing somewhere else.

On the face of it, this system looks rather different from the way words are composed in spoken languages. Spoken words are (a) composed of smaller units (sounds) arranged compositionally, (b) arbitrary and conventional signs, and (c) based on a system of grammatical distinctions. The ASL signs for pronouns, in contrast, look like undecomposable iconic gestures. This is an illusion, however.

For one thing, the "pointing" in an ASL pronominal sign is more arbitrary than it appears at first glance. True, when the reference is to someone present in the conversation, the gesture involves pointing in the direction of that individual. When the reference is to someone not present, however, a more or less arbitrary point in signing space is selected to represent the individual, and pointing to that position constitutes reference to him or her or it.

More interesting, perhaps, is the fact that I make the same gestures, but using a flat hand (oriented toward me, you, or another), to sign "my, your, his or her." Changing the handshape from one with extended index finger to another thus changes the meaning. Or alternatively, I sign the pronoun with an arc movement to indicate "we" (arc from left to right shoulder), "you (plural), they" (an arc with index finger, starting at the spatial location associated with "you" or some third person). The same arc movement made with a flat hand means "our, your (plural), their."

In fact, the ASL pronouns use a pointing finger because it is one of the distinctive handshapes of the language, not because of its iconic function. Likewise, the possessive forms employ another of the canonical handshapes of the language. The signs are indeed compositional: the handshape indicates case (pointing finger for subject or object as opposed to flat hand for possessive). The orientation of the sign indicates person, and the movement indicates singular or plural.

Although the system of ASL pronouns is independent of English, both languages distinguish person. English has first versus second versus third person, while ASL has no formal basis for distinguishing second from third person. The gesture for "you" is simply one instance of a type that indicates any referent visually present in the conversational situation. The ASL person system thus distinguishes only first as opposed to nonfirst person.

English pronouns distinguish three cases: subject, object, and possessive (*I, me, my*). ASL has only two cases: one that we might call "basic" and another for the possessive. This case system is like that of, for instance, Indonesian. English pronouns have two numbers: singular and plural. ASL pronouns have three. Singular and plural are as just described, but there are also dual forms, referring to exactly two individuals, made with a V-handshape. This number system is like that of some spoken languages, such as Slovenian.

ASL also distinguishes between first-person nonsingular forms that are inclusive (you and I, perhaps with some others) and exclusive (some others and I, but not you). The same distinction is present in many spoken languages, including Mokilese (a language of Micronesia that also distinguishes dual from plural forms) and many indigenous languages of North America, such as Kwakw'ala ("Kwakiutl"). English distinguishes three genders in the third person singular (*he, she, it*). ASL does not, but other signed languages such as Taiwanese SL distinguish masculine versus feminine in all three persons. The masculine signs have a fist with erect thumb as handshape, and the feminine forms a handshape with extended pinky finger.

We see, then, that even in an area where we might expect more or less complete iconicity, namely the set of pronouns, signs are compositional, conventional, and based on a system of grammatical categories, just like pronouns in spoken languages. These properties extend to the rest of the signs in the language as well.

It is conventional, in describing the morphology of spoken languages, to distinguish three general categories. One is *compounding,* as in the formation of *hot dog* "sausage sandwich" from *hot* and *dog.* Another is *derivation,* as in the formation of *transmission* from *transmit* plus *-ion.* Finally, we have *inflection,* as in the formation of *dogs* as the plural of *dog,* or of *drove* as the past tense of *drive.* In all of these areas, signed languages display entirely comparable structure.

Compounding is a device for forming new words that is particularly well developed in ASL, as it is in, for instance, Chinese. It involves in both signed and spoken languages the close sequential combination of two (or more) independent signs, typically with an intonation (or rhythmic structure) indicating the unity of the combination. In many instances the combination takes on a meaning that is not simply what would be expected from the meaning of its parts. Compare the 1918 form RED-SLICE "tomato" in Figure 9.5 with English *hot dog* "sausage sandwich." Over time, such compounds may progressively lose their connections to the signs on which they are based, as seen in the evolution of TOMATO or in that of the English *helicopter.* The original connection of *helicopter* with *helico-* "screw-like" and *pter-* "fly" (as in *pterodactyl*) is not recognized by many speakers.

In the area of derivation, just as English has, for instance, systematic ways of forming nouns of various sorts from verbs (*transmit/transmission, bake/baker,* [to] *dance/*[a] *dance*), so has ASL. Some pairs consisting of a verb and a related derived noun are shown in Figure 9.12.

Two areas of inflection have attracted special attention within the study of ASL. The first is the matter of *agreement.* English has rather a limited form of agreement: in virtually all verbs, the only difference is between a form with *-s* (he/she/it *agrees*) and one without (I/you/we/they *agree*). The situation is more complicated in ASL, however, as it is in many spoken languages such as Latin or Georgian.

Of the several different patterns of agreement in ASL verbs, I will mention only two. For the first (and simplest) class (LIKE is an example), the signer simply makes the sign for the verb, in between the pronominal signs for its arguments. For verbs belonging to another class, however, the form of the sign builds in reference to the subject (and to the object, if there is one). A sign like ASK or TELL shows agreement with its argument by being signed along a path that leads from the point in signing space representing its subject to that representing its object.

Another inflectional category that is particularly well developed in ASL (and other signed languages) is that of aspect. English displays very few differences in this regard, primarily the distinction between the simple present or past (*Cassandra sings, sang*) and the present or past progressive (*Cassandra is singing, was singing*). Many of the indigenous languages of

VERB **NOUN**

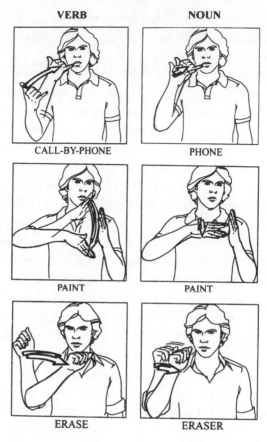

Figure **9.12** Examples of verb-noun derivation in ASL

North America make more distinctions, inflecting a verb to indicate that an action is beginning ("inchoative"), taking place over a long time ("durative"), taking place over and over ("iterative"), and the like. A few of the aspectually different forms of the verb "chop" in Koyukon, an Athabaskan language spoken in Alaska, are given in (1) below. This verb has about fifteen other forms as well. The phonetic symbol [ɬ] in these words indicates a voiceless "l" similar to the sound written *ll* in Welsh.

(1) a. yeeltleɬ "She chopped it once, gave it a chop."
    b. yegheetletl "She chopped it repeatedly."
    c. yootlaaɬ "She chops at it repeatedly."
    d. yenaaltleɬ "She chopped it in two."
    e. neeyeneetlaatl "She chopped it all up."

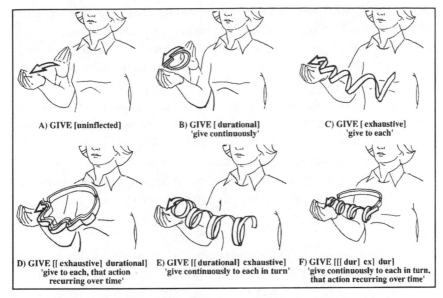

A) GIVE [uninflected]

B) GIVE [durational]
'give continuously'

C) GIVE [exhaustive]
'give to each'

D) GIVE [[exhaustive] durational]
'give to each, that action
recurring over time'

E) GIVE [[durational] exhaustive]
'give continuously to each in turn'

F) GIVE [[[dur] ex] dur]
'give continuously to each in turn,
that action recurring over time'

Figure **9.13** Examples of aspectual morphology in ASL: forms of GIVE

f. yeghedaaltlaatl "She hewed it into a shape."
g. yegheetlaatl "She was chopping it for a while."

Signed languages indicate a number of such distinctions and mark them by modification in the form of the sign (especially its movement). Often more than one of these modifications can be combined to form even more complex aspectual categories. Figure 9.13 illustrates this point with a number of aspectually differentiated forms of the basic sign GIVE.

The aspectual system of ASL is particularly complex and subtle in comparison with that of a language such as English. Degree of mastery of these patterns varies considerably across signers, in a way that tends to correlate with the age at which a signer was first exposed to the language. Those who learned ASL at approximately the normal age for acquiring a first language control it quite fluently, while those who come to signing later in life may never fully master it.

Much of the morphological modification we find in signed languages is of the simultaneous variety. The indication of aspect in Figure 9.13, for example, occurs together with the rest of the sign as a modification of its form. It is comparable to the way English indicates plurality by modifying the basic word in *men, women, mice,* and other "irregular" forms. There are reasons for a manual language to prefer simultaneous to sequential modification, as I show later, but this is not absolute. ASL also contains a few straightforward affixes, signs that are tacked on to another sign to mark a modification of the word's meaning. An example is the sign -ER as in DRIVE-ER, whose structure is parallel to that of its English counterpart *driver*.

ASL structure and that of signed languages in general is fully comparable to spoken languages in the kinds of structure words have. Complex word forms in either modality can be constructed by rule-governed combination of meaningful components, including compounding, derivation, and inflection. These modifications can involve either the addition of one sign to another or a systematic modification of the basic shape of a sign.

Syntax

Most of the attention paid to signed languages has focused on the way individual signs are made: their phonology and morphology. We should not, however, lose sight of the fact that a language is not just a collection of signs—not even a structured collection, in the sense that the signs are related in systematic ways to one another. Natural languages also involve a syntactic system, as we saw in Chapter 8, which allows for the combination of signs into an unbounded range of novel, complex expressions.

Signed languages such as ASL are not merely (structured) collections of signs any more than spoken languages are, no matter how complex their phonology and morphology may be. They have their own syntax, with their own principles of word order, phrasal displacement, and pronoun interpretation. These are not just the rules of English (or French), any more than the syntactic rules of Potawatomi are the same as those of English, although all of English, French, Potawatomi, and ASL conform to the overall constraints of universal grammar that circumscribe the range of possible human language systems.

In some respects, the study of a signed language like ASL broadens our perspective on the nature of language by instantiating possibilities within the scope of Universal Grammar that are hardly ever found in the spoken languages with which we are familiar. By and large, though, ASL syntax offers few surprises.

Among many early misconceptions was the idea that ASL (and by extension, other signed languages) has no grammar: signs are strung along randomly until all of the components of what the speaker has to say have been expressed. This notion probably derives from the fact that simple sentences appear to display considerable variety of word order. Thus, for a simple sentence meaning "John loves Mary," we can find all the orders of signs in (2).

(2) a. JOHN LOVES MARY.
   b. MARY JOHN LOVES.
   c. LOVES MARY JOHN.
   d. JOHN MARY (he) LOVES (her).

These are not simply free variants of one another. The ordering with the initial subject followed by the verb and its object, as in (2a), is in some sense "basic" in that it can occur with no special discourse conditions and no special accompanying nonmanual markers. The others all involve special emphasis and additional markers of "topicalization" or "focus," which I have not indicated in (2). These sentences would be better translated as *Mary, John loves* (but not Betty); (As for who) *loves Mary,* (it's) *John;* and (As for) *John,* (it's) *Mary he loves,* respectively. These variants testify not to a *lack* of syntactic structure in ASL, but rather to the subtlety of the relations between that structure and meaning.

Further, some orders of words are simply ungrammatical. MAN OLD SLEEP-FITFULLY is a grammatical sentence (adjectives can follow their nouns in ASL, as in French), but *OLD SLEEP-FITFULLY MAN is rejected as impossible—just as the corresponding sequence is in English (though not in Latin). JOHN LOVES MARY can *only* mean that John loves Mary, not that Mary loves John. The ASL language has a grammatical organization as determinate as any other. Sentences are associated with meanings in a structure-dependent way, and strings of signs that cannot be assigned a well-formed structure are excluded as ungrammatical.

Important differences between ASL and English show up in the formation of content questions. In English, these require moving the question word to the front of the clause, as in *What did Julie eat yesterday?* In ASL, the question word can be left in the same position as a corresponding nonquestion word (JULIE EAT WHAT YESTERDAY?). Unlike English, the question word in ASL can also appear at the *right* edge of the clause, as in JULIE EAT YESTERDAY WHAT? In this case, we know the question word must have been displaced (not just left in situ), because corresponding nonquestion words cannot appear here: *JULIE EAT YESTERDAY SANDWICH is not grammatical. Furthermore, we cannot have the question word in both places in the same sentence: *JULIE EAT WHAT YESTERDAY WHAT? is also unacceptable.

It appears, therefore, that ASL has a rule which allows question words to be (optionally) displaced not to the left, as in English, but rather to the right. Surprisingly, this issue is the subject of considerable controversy in the literature on the linguistics of ASL. Sentences such as WHAT, JOHN BOUGHT WHAT YESTERDAY? and WHAT, JOHN BOUGHT YESTERDAY WHAT? also occur. These, it appears, are possible only as a result of topicalization, an operation that puts a copy of some phrase at the left edge of the clause. Apparently, simple sentences like *WHAT JOHN BOUGHT? with no "intonation" setting off the topic from the rest of the sentence (which we would expect if question words could be freely displaced to the left) are not possible for native speakers of ASL. The matter remains controversial, although it seems to me the reasons for preferring rightward displacement to leftward are strong.

Question words in subject position can also be displaced to the right, as shown in the examples in (3):

(3) a. WHO SEE JULIE YESTERDAY?
   b. SEE JULIE YESTERDAY WHO?
   c. *WHO SEE JULIE YESTERDAY WHO?

This property of displacing question words to the right is unusual. Spoken languages generally (a) displace their question words to the left, as in English; (b) leave them in situ, as in Mandarin Chinese; or (c) put them in a special "focus" position, typically immediately before the verb, as in Hungarian. ASL is not alone in displaying this unusual property. At least Pakistani, Portuguese, and British Sign Languages (historically not related to one another) also move question words to the right, as does Italian Sign Language (at least as one possibility). Linguists dealing with spoken languages have attempted, without notable success, to provide general principles from which it would follow that only leftward displacement (or displacement to focus, or no displacement at all) can take place. The evidence from signed languages suggests that perhaps they should stop trying.

We have seen that a common structure in ASL makes it possible to topicalize a constituent by putting it at the front of its clause. When that happens, a particular "intonation" (nonmanual marker) accompanies the topicalized phrase, involving raised eyebrows and a slight backward tilt of the head.

The topicalization construction allows us to explore another aspect of ASL syntax: the extent to which it conforms to the same constraints on displacement as those in the syntax of spoken language. In fact, displacement to a position outside one part of a conjoined expression is just as incorrect in ASL as it is in English. The sentences in (4), taken from the literature, show us that the coordinate structure constraint formulated as (34) in Chapter 8 cannot be violated in ASL any more than in English:

(4) a. $_{2nd}$PERSUADE$_{1st}$ BUT PRO$_{1st}$ BUY HOUSE.
    You persuaded me, but I bought the house (anyway).

   b. *HOUSE, $_{2nd}$PERSUADE$_{1st}$ BUT PRO$_{1st}$ BUY.
    *As for the house, you persuaded me, but I bought (anyway).

   c. *MOTHER, PRO$_{1st}$ HIT SISTER HE TATTLE.
    *His mother, I hit my sister and he told [intended: her].

For a further example, this time involving question formation, consider the following situation. In some universities there is no way to major in Linguistics alone. Students must major in "Linguistics and (Psychology, Russian, Computer Science, . . .)." An ASL speaker might thus say something like (5):

(5) ME MAJOR LINGUISTICS SECOND-OF-TWO PSYCHOLOGY.
    I'm majoring in Linguistics and Psychology.

This sentence could also be the answer to a question such as (6), where the question word has been left in place:

(6) YOU MAJOR LINGUISTICS SECOND-OF-TWO WHAT?
    What are you majoring in Linguistics together with?

If the question word is topicalized at the front of the sentence, however, as in (7), the result is as ungrammatical as the corresponding English question:

(7) *WHAT, YOU MAJOR LINGUISTICS SECOND-OF-TWO?
    *What are you majoring in Linguistics and?

Conditions on displacement are also respected in sign, because these appear to be part of the distinctively syntactic system of natural language, rather than following from necessities of meaning or speech(/sign) articulation. The fact that they hold in ASL confirms the claim that ASL syntax really is the syntax of a natural language.

A variety of other arguments support the same conclusion. The conditions under which pronouns can be omitted in ASL are strikingly similar to those that apply in languages like Italian. Basically, when a verb shows overt marking of agreement with some argument, the pronoun is not necessary and can be omitted. When this kind of identification of a referent is absent, however, the pronoun cannot be dropped. This relationship between omission of pronouns and overt marking of agreement appears in a wide range of spoken languages around the world—and also in signed languages such as ASL. Overall, the closer we look, the more the structure of ASL looks like that of a spoken language, except for the irreducible fact that ASL is "spoken" in another modality.

Language and Modality

We see, then, that signed languages display the same principles of organization as spoken ones. Where conditions permit (that is, when a learner—with impaired or normal hearing—has access to the data of native signers at appropriately early stages of development), children learn signed and spoken languages in the same way, in essentially the same sequence, with milestones of development that come at the same ages.

Signs undergo historical change in much the same way as spoken words. Figure 9.14 provides another example, in the development of HOME from an earlier compound EAT-SLEEP, similar to the example of TOMATO in Figure 9.5.

Signers make "speech errors" that deviate from what they mean in terms of the formational parameters of sign, as in the adult signing errors of Figure 9.10. Children learning a signed language may make errors in the parameters of handshape, location, or movement, as in the child's mispronunciation of FLOWER with the movement on the ear instead of the nose. These are entirely comparable to the errors children make in learning the forms of words in spoken languages.

Evidence from brain-damaged signers and from brain-imaging studies suggests that the linguistic aspects of signing are organized in the same areas of the brain as the linguistic aspects of spoken languages, especially in the left cerebral hemisphere including Broca's area and Wernicke's area. Other studies show that the processing of complex (but nonlinguistic) sym-

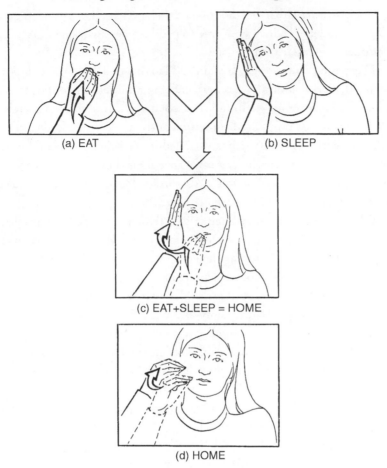

Figure **9.14** Loss of iconicity in historical change
(a) and (b), original iconic signs; (c), formal compound;
(d), modern opaque sign

bolic gestures such as waving goodbye or "thumbs-up" do *not* preferentially engage the same left hemisphere areas as those that seem central to both spoken and signed language.

Although no practical writing system exists for recording ASL, it is nonetheless the basis of a growing literature that includes poetry, drama, and much else. That literature has been an essentially "oral" tradition until the fairly recent introduction of video recording. (Notice, though, that similar limitations did not inhibit epic poets such as Homer prior to the development of writing for spoken languages.) Similarly, in a host of languages for which writing either has not been developed at all or is not in general use, poets and storytellers continue to produce new works and to pass on older ones from their traditions.

There is no doubt, then, that signed languages are complete linguistic systems that arise naturally within Deaf communities. Of course, this does not happen overnight. We know that isolated hearing-impaired children, lacking any signed input at all, spontaneously make up limited collections of "home signs." These have been studied in considerable detail by Susan Goldin-Meadow and her colleagues. These signs show a surprising amount of systematicity and consistency across individuals and cultures, but they are not in themselves at all complete languages.

When such rudimentary linguistic material is provided as input to subsequent generations of language learners, it is quickly capitalized on in the same way that spoken pidgins develop into creoles. It has actually been possible to observe that process in action in the recent formation of Nicaraguan Sign Language. The system developed from simple home signing into a full-fledged language within only a few generations under the attentive (but noninvasive) gaze of a small group of linguists and psychologists.

While recognizing the naturalness of signed languages qua languages, we must not ignore the fact that they display some particular properties that derive from the visual modality. Attention to such effects has been slow to develop, in part because linguists have been at pains to point out the similarities between spoken and signed languages and correspondingly reluctant to point out differences. Now that the basic nature of signed languages is clear, however, research has begun to pay more attention to the ways in which modality affects linguistic structure.

As one example of a property specific to manual-visual languages, signers can switch fluently between the left and the right hands as the dominant hand. In fact, signers do not even really notice whether someone else is signing right-handed or left-handed unless they pay special attention to the question. Signers seemingly translate the raw visual image into the perceptual invariants of sign in some way that is constant under reflection—a cognitive operation that has no apparent literal correlate in speech perception. Such a process of reducing the detail in the signal to what is linguistically relevant is similar to the translation we make in spoken languages to normalize for different speakers, but it is different in its specifics. For those who have learned to sign natively, a "language mode" might well exist in the visual system that parallels the "speech mode" we find in the auditory system.

The visual modality has a much higher bandwidth than the auditory. Thus, a signer can transmit more information at the same time visually than would be possible auditorily. Signing appears to exploit this fact, in that all known signed languages involve considerable elaboration of simultaneous components of individual items. Oral languages generally pile on information by adding more sounds in sequence. Sign, instead, does more things at the same time: for instance, different things with the two hands and the face, or hand movements such as wiggling the fingers simultaneously with the overall movement of a sign.

This difference between simultaneous and sequential elaboration is not absolute. Signed languages use some sequential modifications that are comparable to spoken language prefixes and suffixes. Spoken languages also use some simultaneous modification, as when the vowel of *came* serves simultaneously as part of the lexical identity of the verb and as a marker of the past tense. What is at stake is the relative degree of development of various devices.

The greater use of simultaneous modification in signed languages has a rather interesting consequence. Because the articulators involved in signing (hands and arms) are much larger than those involved in speaking, they are slower to complete their movements, and the production of a single sign takes longer than the pronunciation of a single word. Nonetheless, the possibility of cumulating information in parallel permits signers to transmit content at roughly the same rate as in spoken languages, as a number of studies have confirmed statistically.

If we want to truly understand the nature of signed languages, further analysis of the ways in which the visual modality may shape their structure is surely important. This fact should not distract us, however, from the overall conclusion of earlier parts of this chapter: signed languages such as ASL are just as much products of the human language faculty as the more familiar spoken languages.

A very important consequence follows from this. If an animal could be observed (or taught) to use a system like that of ASL or any other signed language, we could conclude that that animal's cognitive organization was up to the task of human language. It is important to understand just what that means, for there are two sides to the argument. This conclusion would only be warranted if we could show that the animal actually controlled the kind of system we now know to be characteristic of a human signed language, not simply that it could wave its arms about in meaningful fashion. It is to these considerations that we turn in Chapter 10.

10

Language Instruction in the Laboratory

"They say the Doctor talks every animal language there is," said a thick fat man to his wife.

"I don't believe it," answered the woman. "But he's got a kind face."

"It's true, Mother," said a small boy (also very round and fat) who was holding the woman's hand. "I have a friend at school who was taken to see the Puddleby Pantomime. He said it was the most wonderful show he ever saw. The pig is simply marvelous; the duck dances in a ballet skirt and that dog—the middle one, right behind the Doctor now—he takes the part of a pierrot."

"Yes, Willie, but all that doesn't say the man can talk to 'em in their own language," said the woman. "Wonderful things can be done by a good trainer."

"But my friend *saw* him doing it," said the boy. "In the middle of the show the pig's wig began to slip off and the Doctor called to him out of the wings, something in pig language. Because as soon as he heard it the pig put up his front foot and fixed his wig tight."

—Doctor Dolittle's Caravan

Now that we have explored the naturally occurring communication systems of a variety of animals and examined some of the structural characteristics of human languages, it is time to raise a basic question: to what extent do nonhumans (especially other primates) have cognitive abilities that would support the acquisition and use of a human natural language? To put it starkly, how much of human language is uniquely available to humans?

We have already seen that human spoken languages are inaccessible to most other animals for a very simple reason. They lack the requisite apparatus for producing speech. Understanding may well be another issue, as we will discuss especially with respect to Kanzi the bonobo; but neither the vocal tract nor its controlling neurological mechanisms, as these exist in other primates, are adequate to the production of speech. Parrots do not suffer from this limitation, although they employ different means in vocalization. We will therefore conclude this chapter by examining our basic question from a perspective different from that of primate studies.

Apart from Doctor Dolittle's panglossian efforts to develop full language across the animal kingdom (and in some plants as well, in *Doctor Dolittle in the Moon*), research on language abilities that might rival our own has focused on primates, especially on chimpanzees and other higher apes. The first attempts to teach human languages to these animals got virtually nowhere, however. Chimpanzees were brought up by human parents, as normal family members insofar as possible, and unusually intensive efforts were made to teach them language. The result was extreme frustration on the part of both researchers and chimpanzees, but very little linguistic accomplishment for the latter.

One notable case of this sort involved a chimpanzee named Viki. After six years in a human family, Viki had a substantial recognition vocabulary (on the order of thirty-five to forty spoken words), but no command of ways to combine these words. She had a production vocabulary that at its most optimistic could be counted as four recognizable words: *mama, papa, cup,* and (perhaps) *up.* While not a total failure, this project came close; but some reasoned that the difficulty came from the fact that chimpanzees' abilities to produce speech (and perhaps, by extension, to perceive it) were inhibited by purely physiological limitations. We already know that in contrast to parrots, the vocal abilities of chimpanzees and other apes are limited. Their vocal tracts are different enough that they are unable to make most of the sounds that are important in human languages.

We also know that other primates are not at all successful at imitating humans, or at picking up the significance of our gestures. Monkeys are quite incapable of such imitation and interpretation, and apes have only limited capacities. Comparative studies of chimpanzees and human infants suggest that only the humans read intentionality into the actions of others and thereby extract the meaning that may lie behind those actions. Dogs, in contrast, seem to have evolved in a way that makes them quite skilled at reading human communicative signals—although their close relatives, wolves, are not.

It seems reasonable to suggest, therefore, that a good deal of the failure of the earliest ape language experiments was inevitable for these reasons alone, and that those spoken language projects tell us little about the cognitive abilities (or limitations) of nonhumans.

Just as the question of whether apes could learn human language seemed to be coming to a dead end, an alternative approach presented itself. At about the same time linguists were recognizing that signed languages (such as ASL) have all the structural properties of spoken languages, aside from modality. Researchers therefore suggested that it might be worthwhile to try to teach the apes signed languages, on the premise that their control over manual gestures is at least as effective as ours. This approach would provide science with a way to test the notion that animals can in principle learn language, while conducting the experiments in a modality that would avoid the limitations of their vocal apparatus.

Starting in the late 1960s, scientists interested in animals' cognitive capacity for language turned to investigations based on signed languages rather than spoken ones. An animal such as a chimpanzee or a gorilla has hands whose structure and controllability should put these apes well within the articulatory range of signed languages such as ASL.

The nonhuman primate's physical capacity for signed languages may not be perfect, and some physiological differences remain. Gorillas do not have as long a thumb as we do, for example, and it seems impossible for them to make the ASL "W" handshape (thumb contacts pinky, three other fingers extended). But this sort of limitation is minimal and, by and large, a signed language ought to be accessible to an ape in terms of both production and perception, if these are the only factors at stake.

We have seen that signed languages are languages in the full sense of the word—not just collections of iconic gestures, but highly structured systems that display their own phonology, morphology, and syntax. ASL and other signed languages make use of space and spatial relations in distinctive ways that are not available in the medium of sound, but these attributes do not compromise the claim that they are systems of the same fundamental sort as spoken languages, from a cognitive point of view. If an ape really could come to "speak" ASL, we would count it a successful demonstration that human language is within the cognitive capacities of an animal. Recall the caution at the end of Chapter 9, however: such an experiment must show that the animal controls the fundamental linguistic properties of a signed language, not simply that it can gesture meaningfully. Signed languages are much more than gestures, and a valid demonstration of language abilities in another species must be too.

Reaction to these studies on the part of the Deaf community has generally been negative. Many Deaf people see them as demeaning and insulting, based on the notion that while we could never teach a "real" (spoken) language to an ape, it should be possible to do so with the language of the Deaf. To the extent that research looks critically for the significant structural features of ASL in the abilities of the animals, this objection would be misplaced. Unfortunately, the standard adopted all too often is simply that of controlling an inventory of meaningful gestures. In that case, the concerns of ASL speakers are legitimate.

We can blame the lack of positive results in part on deficiencies in some of the experiments. Chimpanzees whose training was in the hands of people largely innocent of the subtleties and complex structure of ASL may have failed to acquire a system anything like the signed language for this reason alone (although hearing-impaired children exposed to rudimentary signing do in fact succeed in developing a much richer language than that of their models). The main reason for the failure of apes to learn the essential properties of a human language appears to be that, as nonhumans, they lack the human language faculty. This is not a value judgment, simply a statement of apparent fact.

Nonetheless, it is important to recognize that we probably have not come close to exploring the limits on the cognitive capacities of animals in the domain of communication. Work with a parrot named Alex (discussed toward the end of this chapter) has produced results more dramatic than anything yet seen in primates—but it is hard to imagine that a bird with a brain so much smaller than those of chimpanzees and other apes is really far more sophisticated cognitively than they are. Limitations of experimental technique, rather than of animal intelligence, therefore may have been responsible for at least some of the limitations of the results of the ape language research.

Classic Ape Language Studies

The experimental projects that tried to teach language to chimpanzees and other higher apes during the 1970s and 1980s got a great deal of attention, both from scientists and from the general public, but they were actually quite limited in number. The studies are expensive, difficult, and time consuming. They require a large and dedicated staff with special training, who must continue to work with the same animal(s) over a long period.

The work is also controversial. For some, the very notion of inducing a quintessentially human ability (language) in an ape is as close to heresy as one can get in a secular age. For others, the failures of previous work make money spent on additional projects a tragic waste of scarce research funding. Criticisms of every sort have made the whole enterprise of "ape language" research a dubious one within the culture of science. So it is perhaps not surprising that no new projects have been initiated for a number of years.

During the heyday of such research, a number of projects explored the linguistic capacities of apes. These are generally known by the name of the animal being studied: Washoe, Nim, Koko, Chantek, Lana, and others. Most were based (in principle) on a sign language as the linguistic system to be taught, though a few (Sarah, Lana, and later Kanzi) used artificial systems involving tokens or keyboards rather than manual gestures.

The first, and probably still the best known, of the early studies is the work done by Allen and Beatrix Gardner with Washoe, and it is there that any discussion of the subject must begin. The perceived accomplishments and limitations of the Washoe project provided the initial stimulus for the work that Herbert Terrace conducted with another well-known research subject, Nim Chimpsky. Terrace's essentially negative conclusions wound up having enormous (no doubt disproportionate) effects on the climate of research on this topic, and subsequent investigators have felt it necessary to discredit Terrace's results as a prerequisite to carrying out work of their own.

Three other projects deal with apes other than chimpanzees. Chantek, an orangutan, has provided interesting hints about the diversity of responses to language training in various primates, but no results that are qualitatively very different from those of the chimpanzee studies. Koko the gorilla has become a sort of folk heroine, and she stands in the popular mind as the canonical instance of "the ape who learned human language." Unfortunately, since this project represents an equally canonical example of how *not* to produce genuinely scientific results from research on the cognitive abilities of other species, we learn next to nothing of substance (though much about research methodology) from what Koko's friend Penny Patterson has written about her supposed abilities.

The studies conducted by Sue Savage-Rumbaugh with the bonobo Kanzi are totally different from those of Patterson. In addition to her earlier work with the chimpanzees Sherman and Austin, Savage-Rumbaugh has documented Kanzi's behavior and ability in great detail over a long period, and as a result a meaningful and very important record is available to consider and evaluate. It is Kanzi who presents the most serious and genuine challenge to those who doubt the linguistic capacities of any nonhuman animal. In the end, one comes away with the conclusion that Kanzi displays fascinating cognitive abilities not previously seen in any nonhuman primate—while still falling well short of what one would have to require of an animal who has truly acquired the structural core of a human language.

When we read on the science pages of the *New York Times* or elsewhere that "apes have learned to communicate in a human language, ASL," the evidence comes almost exclusively from the studies enumerated above. Such a conclusion would be incredibly interesting if it were correct, but we need to be critical and ask the hard questions. These include (among many others): How much system is there to what the apes in these experiments have learned? Have they actually learned ASL, a naturally occurring human (manual) language? If not, to what extent does what they *have* learned display the essential linguistic properties that could convince us that (like ASL) it is a natural language?

Washoe

The Gardners obtained Washoe, a wild-born female chimpanzee, at an age somewhere between 8 and 14 months. In June 1966 they brought her to a trailer in their backyard in Nevada, where their initial idea was simply to bring the animal up with sign being spoken around her, in the hope that she would learn it naturally as a human child would. In the beginning Washoe did not seem to be making much progress, or indeed to be paying any attention to the signing. In retrospect, we can see that this is not remarkable, since we now know that chimpanzees are rather poor at interpreting human gestures of any sort, even basic pointing, as significant.

Because Washoe was not progressing on her own, the Gardners modified their procedure: instead of just making signs and hoping she would catch on, they would show her an object and then mold her hands into the position for a corresponding sign. If she subsequently made the gesture on her own, she was rewarded. This theme is worth our attention: virtually all of the "utterances" we find reported in these projects are requests (directly or indirectly) for gratification, such as a preferred food, tickling, play, and the like.

The molding technique worked. Before long Washoe could produce a fair number of signs, and she had even learned a few from observation alone, without molding. The Gardners were trying to be careful and wanted to be sure that they did not ascribe a sign to Washoe without solid evidence. They established as a criterion that they would not count a sign as "learned" until it had been produced spontaneously (that is, not directly after seeing the same sign from a trainer) on fifteen consecutive days. That was easy enough at the beginning, but as Washoe learned more and more signs, she soon had no occasion to make most of them on any given day. Accordingly, Washoe's training came to include a lot of vocabulary testing, a great deal of "What's this?" activity.

By the time Washoe was 51 months old, she had acquired some 132 signs by this criterion. The project ended for her at the age of 60 months, at which point she had 160 signs. In 1970 she was "retired" to the Institute for Primate Studies at the University of Oklahoma. Roger Fouts has written in very moving terms about Washoe, her life with the Gardners, and much later investigation of his own. Interesting as the anecdotal reports of Washoe's later years may be, they do not provide data of the sort that would motivate a major revision of the conclusions from other work about the strictly linguistic abilities of chimpanzees or other apes.

Between 1972 and 1976 the Gardners brought several other chimpanzees into their laboratory. Moja, Pili, Tatu, and Dar were each adopted shortly after birth and raised with human sign language trainers much as Washoe had been. The results of these studies have elicited far less comment than the work with Washoe. Since the results were not significantly different, I mention them below only where they provide specific evidence not available from Washoe.

Washoe's signs were fairly general. They were learned with respect to a particular exemplar, of course (a specific dog as the occasion of learning to sign DOG, for example), but were quickly used in broader ways. For instance, a sign that Washoe learned early was interpreted by the Gardners as MORE. The ASL sign MORE involves bringing the two hands together so that the fingertips touch. Washoe, however, made her sign with palms facing her (only one of many instances in which her signs differed in major ways from those of the language she was supposedly acquiring). Washoe's MORE was first used together with TICKLE, and then extended to other requests.

The ASL sign for OPEN is flat hands, palms out, index finger edges together, swinging out so the two palms face. Washoe used a different "index" handshape, with hands together face down which then separated and rotated upward. Initially Washoe used this sign with three specific doors; she then extended it to all doors, containers, faucets, and the like, which goes well beyond simple imitation. The human signers in Washoe's environment did not use OPEN for a faucet.

On the other hand, OPEN is a sign which, like many others in ASL, incorporates its referent in the form of different handshapes that serve as "classifiers" for the object that opens. As a result, OPEN DOOR is distinct from OPEN WINDOW, or from OPEN in general. This aspect of structure (classifiers) is prominent in a number of signed languages that have been studied, but was never reported in the signing of Washoe—or any other ape.

The reason, at least in this instance, is not hard to find. None of Washoe's trainers controlled ASL well enough to use classifiers productively in their signing to her. Without having demonstrated command of this aspect of the natural language ASL, an animal cannot be said to have learned the language. The fault may not be Washoe's (although human children do generalize classifier usage from extremely limited input), but this is not the place to give her the benefit of the doubt.

How do we know Washoe was actually making signs, not just gesturing? The Gardners allowed a lot of sloppiness in her signs, on the grounds that her hands were shaped differently from human hands. In studies of this sort, if the observer knows what the answer is and is willing to accept rather inaccurate renditions of it, chances are all too good that the data will be overinterpreted. To prevent this, the Gardners did a series of double-blind tests, where the experimenters coding the response could not see the object the chimpanzee was supposed to identify.

Under these conditions, the observers' interpretations of Washoe's responses corresponded to the object she was supposed to be identifying about 60 percent of the time. Later experiments with Tatu and Dar produced about 70 percent and 52 percent correct answers. It is hard to determine the variation from chance here, because we do not know the size of the set of possible answers on any given trial. These experiments focused on whether the animal would produce a result of the appropriate class (as discussed below); the question of whether the answers were factually correct was secondary.

It would be valuable to know whether Washoe ever signed about things that were not present in the immediate environment. If she did, it would indicate some independence of the sign and the referent. Washoe did make signs for food that was not present (generally as a request), or actions that were not being performed (tickling). In one famous incident she heard a dog bark and made a sign for DOG. In ASL DOG is made with the right hand patting the knee while fingers snap; Washoe's sign involved a hand moving down to the side of the leg. The dog was not visually present, but it *was* auditorily present. We would need a large corpus (say, a record of all of her signing for a day or more) in order to know how much of her production was spontaneous, what kind of context was present in each case, and so forth. In fact, the only records available consist of individual isolated incidents, together with a summary of vocabulary.

What evidence do we have for linguistic structure that goes beyond the production of individual signs? Washoe often produced multiple signs in sequence, but it is tricky to know when to treat such sequences as complex combinations representing a single concept, and when to see them merely as one sign after another. Some combinations of signs do seem to have occurred, and some of these were evidently novel (in the sense of not having been present as such in the signed input Washoe saw from her human companions).

Reported examples include GIVE TICKLE, GO SWEET, OPEN FLOWER, although the last two would actually be ungrammatical in ASL. In that language, the signer would introduce the candy or the flower and assign it a location in space, then make the verb sign with an orientation to that location. We can see that Washoe's combinations were not just imitations, which attests to the creativity underlying their production. However, they make it clear that basic features of ASL (the system of spatial deixis and the indication of agreement based on it) were not controlled by the chimpanzee. Again, this may be a result of the limited knowledge her trainers had of ASL, but that does not lessen the importance of the point.

Other combinations were emphasizers: OPEN HURRY. By far the most famous of Washoe's signed combinations was her production of the sequence WATER BIRD on seeing a swan. Much has been made of the apparent creativity of this novel compound, but we would need to know a great deal about the circumstances of its production before we could construe it in that way, as I will have occasion to observe below.

Some combinations included (apparently) three, four, or more signs, and there is no reason to doubt that sequences at least that complex were possible. The manner in which the Gardners recorded and analyzed their data, however, makes it impossible to decide how much structure, if any, these sequences had.

Overall, what kind of structure *should* we attribute to the sequences of signs Washoe produced? A significant problem for the Gardners was that not much was known about ASL structure at the time, so they had little guidance with regard to what they should be looking for. Nor were they themselves particularly fluent signers. In fact, much of the time it appears that they and their assistants were not actually using ASL syntax. Most of what they produced was English, with signs substituted for words.

This "signed English" is one way that human deaf children are sometimes taught. Quite a bit of research now shows, however, that this kind of system (with signs substituted for the meaningful units of spoken English) is not actually learnable in the way a natural language is. Children exposed to such input either fail entirely to generalize within this system, or else creolize it and turn it into something else that is more like ASL. This was clearly a major methodological problem with the Washoe project. However, since that is what the input was from which Washoe was expected to learn, we have to ask how to assess her success.

To support the claim that Washoe's signing incorporated some grammatical structure, or at least some appreciation of such structure, the Gardners asked her a series of content questions (WHAT'S THAT? WHO'S THAT? WHOSE IS THAT? WHAT COLOR IS THAT? WHERE WE GO? WHERE SHOW? WHAT NOW? WHAT WANT?). The hope was that Washoe would consistently give answers to WHAT questions that would consist of common nouns, answer WHO questions with proper names, and so on. They had the experimenters ask her these questions several times a day. They collected answers until they had fifty responses to each question, and then coded the type of answer.

Mostly, Washoe did well on questions about WHAT, WHO, WHAT COLOR, and WHOSE (noun). *Where* questions, however, yielded a much higher number of inappropriate answers. When the experiment was performed with Tatu and Dar, the only questions considered were of the type WHAT, WHOSE, WHAT COLOR, and WHAT MATERIAL. The hope was to show that the animals had a system of distinct grammatical categories for their signs, but this is a peculiar interpretation to assign to what was actually tested. The categories were at least as plausibly based on semantics as on grammar, so the results tell us little if anything about grammatical understanding.

In fact, the situation is even worse than that. If Washoe was asked WHAT THAT? when shown a dog, and she responded GRAPE, she got full credit, because GRAPE is a common noun; and if asked WHAT COLOR THAT? about the same dog, she could receive full credit for ORANGE. As long as she got the right category, she did not have to give any evidence that she was answering a question about the relevant object.

Further, many answers involved more than one sign, and the sequences were systematically simplified when recorded by eliminating any and all repetition. Thus, in response to WHAT WANT? Washoe might produce YOU ME YOU OUT ME, which would then be truncated to YOU ME OUT and coded as WE OUT. The ultimate result looks like a plausible answer, but we cannot tell how much of this utterance Washoe might have intended as responsive to the question, or even how much of the recorded utterance was actually Washoe's as opposed to the interpretation of the experimenter. Since all we can see are the reduced codings, we have no idea how much redundancy and simplification were involved and subsequently cleaned up by the coding system.

So it becomes even more problematic to interpret her longer utterances as genuinely syntactic. The sequence YOU ME YOU TICKLE ME YOU TICKLE TICKLE ME YOU would get coded as YOU TICKLE ME, a result that looks much more like language than the uninterpreted original. The Gardners were explicit about the kinds of reduction they made in coding the animals' utterances, but it would still be necessary to see the originals in order to evaluate their character as language.

What about the combinations Washoe produced that were genuinely novel? We have no real way of telling that they were in fact combinations. WATER BIRD could have been a case where Washoe was asked WHAT THAT? and first attended to the water, then noticed the swan, and signed BIRD. They might be two utterances, not a combination.

It is not that these matters are undecidable in principle, only that the evidence that would help us decide is not available. In English, when we put two nouns together in a compound, they are given a particular distinctive pattern of stress. Contrast *bláckbìrd* (a compound) and *blàck bírd* (a phrase). ASL also has stress (realized by force of movement, not of course by loudness or pitch), and ASL compounds involve a shift of stress to the second element. The first sign in a compound is reduced: for instance, RIVER is a combination WATER FLOW with the first sign reduced, and GRASS is similarly like GREEN GROW with reduction of the sign GREEN.

A clear way of marking compounds therefore exists in ASL, but we have no evidence that Washoe did anything like it—or even that the Gardners would have known to look for it, since they were not signers themselves, and the indications are subtle. Without a lot more evidence, we simply do not know how to interpret these sequences, and we certainly do not know that they were intended by Washoe as complex sign combinations.

This conclusion brings up some pervasive problems with the early experiments. On the one hand, the experimenters were in many ways pioneers, so there are many matters on which we would like, in retrospect, to have much more data (and data of different sorts) than was actually collected. But there is a much less benign side of the "missing data" problem. The early experimenters did not make much useful data available for study by others. By and large, they presented only their conclusions, some summary counts, and a few appealing anecdotes, but not the data on which the conclusions were based, or enough material to allow someone else to judge the representativeness of the anecdotes. Early criticism of the work of the Gardners and others seems to have produced in them an extremely protective and defensive attitude toward their data, and that is just not the way science is done.

The Washoe project suggested strongly that it is possible to teach chimpanzees a substantial vocabulary of arbitrary signs, in the form of manual gestures with an associated meaning that is at best only partially related to the form of the gesture itself. Little or no evidence exists for any linguistic structure beyond this, and certainly none for full (or even substantial) command of a human language.

I should include another cautionary note about the individual signs. Not many of Washoe's signs were very much like the ASL signs she was supposedly learning. Her HURRY was a shaking of the wrist, while in ASL HURRY is signed with both hands in a specific handshape ("H"), palms facing, moving alternately up and down. Washoe's HURRY sign seems to have been quite unlike the ASL form. It is, however, remarkably similar to a natural gesture made by chimpanzees in the wild, identified by Jane Goodall as linked with general excitement. Not all of Washoe's signs have such obvious sources in the animal's natural gestural system, but it is crucial to establish these precedents in order to avoid inflating the inventory of "signs" we appear to have found.

Nim Chimpsky

Washoe was the first chimpanzee to undergo something like systematic training in "sign language." I have already raised some questions about whether that was actually what she was taught, and about what she learned in the way of signs—and I will return to those matters later—but that was the premise. Certainly the initial reports that came out of the Washoe project tended to make people think that a natural signed language (ASL) was what Washoe learned.

In early 1973, Herbert Terrace—a psychologist of basically behaviorist inclinations at the time—started another project, whose goal was to extend the results of the work with Washoe. As a behaviorist, Terrace was interested in the extent to which language could be taught to a chimpanzee. If language learning is merely the acquisition of a conditioned behavior, it ought to be accessible to a chimpanzee. Beyond that, he was interested in being able to talk with the animal, to find out how chimpanzees see the world. On one of the early public television programs in the Nova series, he advanced the notion that he would take Nim to Africa and use him as an interpreter with other chimpanzees.

Apart from these rather nebulous, global goals, Terrace wanted to explore the issue of how much linguistic structure a chimpanzee could acquire. Although reports from the Gardners suggested that Washoe produced not just signs, but combinations of signs, it was difficult to tell how reasonable it was to attribute linguistic structure to those combinations. Terrace wanted to ask: "Can an Ape Create a Sentence?" (in the words of the title of his well-known 1979 article in *Science*).

Terrace's bias at the outset was toward a favorable result. B. F. Skinner had proposed in 1958 that language was simply "verbal behavior" and that it was learned through the same sort of reinforcement regime as all other associative behavior. Noam Chomsky had argued that this theory was completely inadequate, and that we needed to assume a much richer innate system, especially to account for language acquisition. Terrace believed that Chomsky's refutation of Skinner was overstated and excessively *a prioristic.* Other influential psychologists (Roger Brown, for instance) also doubted that an ape could control syntax, but this opinion was based on at least some rudimentary data, as opposed to mere philosophical predisposition. Terrace hoped to resolve what he thought of as a real empirical issue.

Nim Chimpsky was a captive-born two-week-old chimpanzee when the project began. He was initially reared with a human family: that of a former student of Terrace's, Stephanie LaFarge, who had had a first try at raising a chimpanzee a few years earlier without attempting language. LaFarge knew some ASL, though she is not a Deaf (or native) signer. The premise was to raise the chimpanzee as a human infant is raised. At the age of 18 months, Nim moved from the LaFarge household in New York City to an upstate mansion owned by Columbia University.

Systematic language training had begun at 9 months. Every weekday Nim spent about five hours in a specially designed classroom at Columbia, where a great deal of recording and videotaping went on. Trainers (of whom there were many, though some, like Laura Petitto, were associated with the project over rather long periods) were supposed to sign with Nim, although for the most part they were not fluent signers either. They whispered their interpretation of Nim's signing into a tape recorder and prepared transcriptions later. A number of transcriptions of videotapes of Nim's signing at home were made as well.

The data collected in this project have largely been made available, and constitute essentially the only corpus of signing-ape data from any of the early projects. This is a rather interesting fact. As we have seen, most of the other projects adopted a rather defensive tone from the beginning, with a reluctance to let other researchers see the raw data on which their claims were based. The Gardners actually threatened to sue Terrace for the analysis he made of Washoe data derived from the Nova films.

As with Washoe, the main way Nim learned signs was by molding: the teacher would actively form Nim's hands into the desired sign. Some few signs were acquired by imitation, once the vocabulary had begun to develop. Nim's first sign (DRINK) appeared at 4 months. By the end of the project, when Nim was 3 years 8 months old, he had acquired a vocabulary of about 125 signs. He signed quite a bit, and a corpus of about 20,000 multi-sign utterances (by no means all different!) recorded during one period of two years is available for examination.

The early ape language projects often compared the abilities of the animals with those of young children at the first stages of language learning. At the very beginning, when children are producing only single words, it is hard to attribute sophisticated grammatical structure to them—and correspondingly easy to find an analogy in the behavior of an animal that produces isolated signs. Even when children enter the "two-word" stage, and begin to produce meaningful combinations, it is difficult to know how much knowledge of structure beyond mere vocabulary to see behind their utterances. Accordingly, it is difficult to refute directly a claim that chimpanzees producing sequences of signs are doing just about the same thing as children at this point. However, a growing body of evidence supports the conclusion that children have a more sophisticated understanding of grammatical structure than might be immediately evident from their productions.

The path of language acquisition in the child after the very first word combinations are produced is somewhat different from what we observe in chimpanzees such as Nim. A common (if extremely coarse) measure of this development is the child's (or chimpanzee's) Mean Length of Utterance (MLU), an index of the average length of utterances in numbers of meaningful units. From the data recorded in the Nim project, we can see that while he continued to produce sequences of signs, his MLU did not really increase. During the last year and a half of the project it was around 1.1 to 1.6, rather than rising into the 2–3 range, as we would expect for human children at (supposedly) comparable stages of development. The strong implication is that human children have a much more structured framework into which to integrate multiple word combinations than chimpanzees do.

Sign Combinations

Let us look at multisign combinations a bit more closely, to see how they might be interpreted in Nim's productions, or Washoe's, or those of any other nonhuman animal. Given a sequence of gestures that we can interpret as a two-sign utterance, there are a variety of stories we could tell about it and we need to ask how to distinguish them from one another.

One possibility is that we are simply observing superficially "complex" signs without significant internal structure. The chimpanzee has learned that certain sequences of signs have a holistically determined effect, although the components into which we might break them have no independent significance for the animal. For instance, what the experimenter analyzes as TICKLE NIM might be a complex action designed to elicit tickling, not the combination of independent ideas "tickle" and "Nim."

Another possibility is what we might refer to as the "semantic soup" theory. On this view, the chimpanzee has a lot going on in his head at a particular moment. Some of these thoughts correspond to signs he knows, and he produces the corresponding gestures. The signs that emerge reflect his ideas, but with no particular organization apart from general contextual salience. They are organized, but purely in terms of conceptual simultaneity.

Still another possibility is that the sequences we observe are formed by a system based on what Pinker refers to as "word chains" (mentioned in Chapter 8 as a finite state device). The signs are independently significant, but their order is determined as a fact about independent lexical items. For any given word, the animal has some knowledge of which words might come next, but nothing more. Thus, in any utterance where both "you" and "me" occur, Nim reportedly preferred to have "you" come first.

Finally, we might be seeing the workings of true hierarchical syntax: principles based on a classification of signs into grammatical categories, organized into constituents of various types; utterances with the form NP VP, where anything that is a possible NP comes first, and so on. And since constituents can contain other constituents, potentially of the same type, in principle this kind of structure has no upper bound of complexity. That is, it is *recursive*, although of course practical constraints on length that may be imposed by memory and other factors.

All of the above are logically possible accounts of what underlies the production of a multisign utterance by a chimpanzee (or the multiword utterance of a child). We need a way to distinguish among them; but in regard to, say, Nim, the evidence we have is really only the relative order of the signs as produced. When it comes to Washoe, the method of coding multisign utterances removes much information even about order.

With animals, the most powerful tools for exploring the degree of hierarchical, constituent-based syntactic structure cannot really be applied. That is because no chimpanzee has gotten to a point where it would be possible to ask, for instance, how to form the question corresponding to "The boy who is tall is tickling Nim." Children can tell us that this should be "Is *the boy who is tall* tickling Nim?" and thus confirm that *the boy who is tall* is a single noun-phrase constituent in their grammar (just as the single word *Nim* is), but there is as yet no way of asking anything comparable of a nonhuman language subject.

So we are left with what we can extract from the available evidence in the way of regularities of sign ordering. When we look at collections of chimpanzee utterances, seemingly the tendencies in ordering are only that: tendencies. That is, we do not find the fairly strict regularities that might be attributed to rules.

When confronted with the apparent absence of genuine rule-governed principles of ordering in the data from their chimpanzee subjects, the Gardners, Roger Fouts, and others responded in an interesting way. They argued that their chimpanzees were learning ASL, not English, and that while English has strict word order, ASL does not. The problem with this argument is that ASL has other aspects of grammatical structure that are relevant.

The basic order of sentence constituents is preferentially S(ubject)-V(erb)-O(bject), although OVS order is also possible where no ambiguity results: thus, both MARY READ BOOK and BOOK READ MARY can occur, with the same basic meaning. However, many ASL verbs are inflected to show who does what to whom: JOHN LOOK-AT MARY is signed with an orientation from a point in space representing JOHN to a point representing MARY. When a verb agrees with its arguments in this way, the order of overt noun-phrase expressions JOHN, MARY (if these are present at all, which they need not be) follows principles of discourse salience, rather than syntactic relations.

We have no evidence that the apes in any of the experimental projects ever do any of this when signing. Their ordering possibilities do not seem to be constrained by possibilities of misinterpretation, and they do not inflect signs to agree with their arguments in the way ASL signers do.

This is not surprising, actually; because most of the teachers Washoe and Nim had were not fluent signers, they did not produce "real" ASL any more than their models had. What they produced was a sort of pidgin signed English: English sentences (with words replaced by signs) with English order—though generally without grammatical markers for categories like tense and the much more limited form of agreement that English shows. Grammatical relations were indicated by regularities of order, but there is no reason to believe the chimpanzees ever picked up on this, and of course they had virtually no evidence for the grammatical mechanisms of true ASL.

Despite the intentions of the experimenters, the evidence from which their chimpanzees were supposed to learn their language was based on significant ordering of signs, not on the more order-independent mechanisms of ASL. We cannot therefore conclude that order is irrelevant in this language, and we are left with the question of just how much structure is implied by the order we find.

Structure in Nim's Signing

Terrace undertook an analysis of Nim's signing to explore these issues. Among the various possibilities suggested above, he could immediately exclude the one in which multisign combinations have no internal structure such that sequences of signs are holistic units, on the basis of the number of different token combinations Nim produced. These included something over 2,700 distinct types of combination of two- and three-sign sequences, arguably far too many for the animal to have memorized as distinct units.

Similarly, the theory that sequences derive entirely from the ordering preferences of individual items, along the lines of the word-chain model, seems excluded. Even though some items have strong preferences (for instance, MORE is generally initial), the preferences for some sequences over others cannot be derived from the independent ordering probabilities of the individual signs in statistical terms.

We are left with the possibility of significant structure, and Terrace offers one argument for a structural interpretation. The majority of Nim's (and Washoe's) multisign utterances can be classified into a small number of categories such as "agent-action," "action-object," "modifier-modified," and a few others. These are, of course, the kinds of semantic relations that are present in simple syntactically structured utterances in human languages, and perhaps Nim controlled a similar system.

But why, one must ask, does this constitute an argument for anything beyond what I have called the semantic soup theory? Perhaps Nim's internal state on an occasion when he produced a sequence of signs included an awareness of something that was going on (or that he wanted), and also of someone or some thing that was (or should have been) the agent or the object of that action. That still does not mean that the signed utterance Nim produced codes the relation among these ideas, in addition to the various components individually. To demonstrate this, one would have to show at a minimum that the orderings (of, for instance, the agent and the action) were consistent, and not derivable from some much simpler principle such as contextual salience. And in some cases (action-object, object-beneficiary), both orders of the signs involved occur with about equal frequency in the data on Nim's signing.

Nim's multisign utterances, similar to those of Washoe (to the extent we can determine this), display a marked difference from those of human children. As Nim signs more and his utterances get longer, they do not get more informative. Nim tends to produce repetitions, of the GIVE ORANGE ME GIVE EAT ORANGE ME EAT ORANGE GIVE ME EAT ORANGE GIVE ME YOU variety—many signs long, it is true, but containing only the information of "you give me (an) orange (to) eat." Human children essentially never do this, though they certainly repeat whole utterances, or even individual words, for emphasis.

In 1979 Terrace and his colleagues published a paper in the journal *Science* that had a tremendous effect on the scientific community involved in ape language studies. Their work concluded that, when one explores the discourse context of utterances, Nim's utterances rather directly reflected the teacher's signing. That is, many multisign utterances on the chimpanzee's part were actually initiated by the teacher, and involved signs that occurred immediately before in the teacher's utterance. As a result, the amount of signing where we can say that the structure is the product of the chimpanzee's control of the language is really quite small, and it provides little or no evidence for real structural regularities.

Notice that Terrace and his colleagues did not say that chimpanzees do not sign spontaneously, although some critics accused them of claiming this. Nim and Washoe clearly did make gestures when they wanted things— and perhaps for other purposes as well, though this is much less certain.

But the fact that so much of the potential evidence for syntactic structure came from prompted utterances that were at least partly repetitions of what the teacher had just said greatly reduces the evidence for syntax. Terrace showed that to the extent evidence was available (from videos extracted from the Nova presentations), close analysis of the productions of other signing apes (Washoe, Koko) showed the same repetition of teacher utterances.

While Terrace's analysis of the signing patterns of Nim and the earlier language-trained apes was carefully and accurately done, the phenomenon he uncovered may be due at least in part to the training situation in which the animals were recorded. Several years after Nim was retired from the project bearing his name and returned to the Institute for Primate Studies in Oklahoma where he had been born, another team of researchers visited him and recorded a series of interactions. His behavior when they drilled him on naming items in the way much of his earlier training had proceeded was entirely comparable to what Terrace and his colleagues recorded in their transcripts. Nim obviously did not like this activity and quickly became hostile; the session was ended when he bit the investigator. In a more relaxed and conversational interaction, however, the transcript of his signing suggests more spontaneity, and less repetition.

Under these conditions, Nim's signing was still almost exclusively related to requests for food, toys, and pleasurable activities. There is also no further evidence for structured sign combinations of a sort that would suggest syntactic organization. Still, his conversational behavior was qualitatively quite different from that in the training and testing situation. A full appreciation of what an animal can do with the communicative tools acquired in training seems to require a more creative approach than was characteristic of most of the classic ape language studies.

Terrace's central conclusion was that there was no evidence in the ape language research for syntactic abilities of the sort crucial to human language. We have no reason to question that result, even in light of the evidence that Nim had greater conversational abilities than he showed in the Columbia study. In this regard, it is ironic to note the subtitle of Terrace's book *Nim:* "A Chimpanzee Who Learned Sign Language." This subtitle was apparently introduced by the publisher, despite the much more modest (indeed, almost opposite) conclusions of the book. Most of those who paid attention to Terrace's volume interpreted the results of project Nim as showing that the effort to teach language to nonhuman primates had

failed. Funding for further research into the question became much harder to find.

After the appearance of the reports on Nim, researchers engaged in the other ape language projects became more defensive and retreated to unsubstantiated claims that Nim was an unfortunate choice of subject, or had too many teachers (thus making him more dependent on those teachers because of emotional deprivation), and the like. Of course, what Terrace had shown was that syntax could not be attributed to chimpanzees—not that they had not acquired incredibly interesting abilities. What they had learned was not human language, perhaps, but it was hardly negligible.

Projects Involving Other Apes

While chimpanzees are often said to be the apes that are closest genetically to humans, and thus the most obvious candidates for language-learning experiments, the other great apes (orangutans, gorillas, and bonobos) have also figured in this work. The number of projects involving nonchimpanzees is quite small, but two respond explicitly to the criticisms of the Nim project, so I mention them first. One involved an orangutan, Chantek, and the other a gorilla, Koko. (I discuss work with bonobos, especially Kanzi, separately.)

Chantek

Orangutans are the only Asian great apes, and they have not been the focus in as many studies of cognition as their African relatives. Chantek is the only orangutan who has been studied with respect to language ability, though his trainer Lyn Miles says explicitly that "the goal of this research was not to demonstrate whether or not Chantek had acquired 'language'" but rather "on a developmental perspective that seeks to identify the cognitive and communicative processes that might underlie language development." She concludes that Chantek did indeed develop an ability to use manual gestures (signs) in a referential way—an important result in its own right, independent of more controversial claims about full human language.

Chantek was born in captivity at the Yerkes Regional Primate Research Center in Georgia in 1977. At the age of 9 months, he was moved to the University of Tennessee, where Miles worked with him until 1986. Unlike Nim, he was raised in a fairly relaxed environment. There were no trips to a specially designed classroom for sign lessons; rather, signing was taught in his customary home cage. "Class" generally consisted of simply being around trainers who signed to him about what was going on in the environment. Again in contrast to other studies, his training involved very little vocabulary drill, and more emphasis on the utility of signing to get what he wanted or liked.

At the outset, Chantek was introduced to signs through the technique of molding, but eventually he began to pick up signs by imitation. In reporting her results, Miles uses the same strict criteria for "knowing" a sign as the Gardners, and Chantek's rate of vocabulary growth was about the same as Washoe's and Nim's. This result makes it clear that vocabulary drills are not necessary to get apes to learn signs, at least not after they have learned the first few. Miles also provides us with an indication of the number of different signs used every day, showing how this increased over time. We still do not have anything like a full record of Chantek's utterances, but this is information of a type that is not available for most other studies. We can see that Chantek continued to use old signs while learning new ones.

In reporting on her work with Chantek, Miles explicitly responds to Terrace's observation about the role of imitation in the signing of other apes. While upward of 30 to 40 percent of Nim's utterances were direct imitations of his trainers, she claims that only 3 to 4 percent of Chantek's were. About 8 percent of Nim's utterances were spontaneous, as opposed to 37 percent of Chantek's. That is, Chantek was much more likely to start a conversation, or just to start signing without prompting, whereas most of Nim's signing was in response to prompting.

Like the others, Chantek apparently began to produce multisign combinations after learning only a few signs. Miles argues that this process was not just the kind of repetition seen in Washoe and Nim, but she does not provide any lists of multisign utterances, statistics on the ratio of combinations with and without repetition, and so on, so the record is very hard to evaluate.

Miles is also quite explicit that what Chantek was exposed to was not ASL, but rather Signed English. His input had English word order, with signs substituted for words, and all grammatical markers (agreement and tense endings, articles) omitted. As a result, of course, he did not come to control ASL syntax; but we have no evidence that he controlled English syntax either.

Since no one claimed that Chantek "learned language," the importance of this work lies elsewhere. First, we note that Chantek acquired a vocabulary of about 140 signs, showing that the ability to learn this kind of communicative system is not limited to chimpanzees (and humans). As with the other apes, his gestures differed in many ways from those of actual signs in ASL — Chantek apparently liked to sign with his feet, for instance — but there is little doubt that he did develop a significant set of mostly arbitrary meaningful gestures, which he achieved with minimal explicit training.

Chantek also displayed a number of indications that his signs had genuinely referential values for him, rather than being simple context-dependent gestures. These included his signing for objects that were not present in the situation (or at least not visible), as well as extending the reference of a sign to other things that were similar but not identical to its original sense. The sign for DOG came to be used for a variety of dogs, pictures of dogs, and a number of similar animals, BEARD was used for hair in general, and many other examples occurred. Since there is no evidence that orangutans (or any other apes) use arbitrary signs in a referential way in nature, the demonstration that they can nonetheless develop such communicative skills in the laboratory is of considerable interest.

Koko

Chantek got relatively little attention in comparison with Washoe or Nim — or with another project, that of Francine (Penny) Patterson's gorilla. Koko has been consistently presented as the ape who "really" learned sign language, and who uses it the way humans do — swearing, using metaphors, telling jokes, making puns. But make no mistake, we have nothing but Patterson's word for any of this. She has not produced anything for anyone to look at except summaries (lists of signs, charts of rate of vocabulary growth), and isolated stories. She says that she has kept systematic records, but no one else has been able to study them. This project is the best illustration imaginable of the adage that "the plural of 'anecdote' is not 'data.'"

Koko was a year old when Patterson began working with her in 1972. Initially she was trained just like Washoe and Chantek, with molding of signs. Patterson also spoke aloud while signing, and it is reasonably clear that Koko's input consisted of a sort of pidgin Signed English rather than real ASL. Like Chantek, Koko caught on after a while and began to imitate. Patterson used a slightly less stringent criterion for learning than the Gardners, but also did not do a lot of artificial drilling on vocabulary. By the age of 3½, Koko reportedly had acquired about 100 signs, and by age 5 almost 250. On double-blind object recognition tests, she scored around 60 percent correct, roughly the same as Washoe and the other chimpanzees in the Gardners' studies.

Although limited amounts of summarized information about Koko's signing were published in the early years of the project, none of it included the kind of raw data scientists would need to come to a reasoned assessment of her abilities. Patterson says that she keeps detailed records and transcripts of Koko's signing, that she videotapes extended sessions, and so on, but none of this material has ever been available to outside scientists for analysis and assessment.

Since 1981, information about Koko has come only in forms such as Nova or National Geographic television features, stories in the press, children's books, Internet chat sessions (mediated by Patterson as interpreter and translator in both directions), and the ongoing public relations activities of the "Gorilla Foundation" (currently soliciting funds to enable Koko and her entourage to move to Maui). We are told a great deal about how clever and articulate Koko is, but in the absence of evidence it is impossible to evaluate those claims. And what we do have does not inspire great confidence. Here is dialogue from a Nova program (filmed ten years after the start of the project), with translations as provided for Koko's and Patterson's signing:

Koko: YOU KOKO LOVE DO KNEE YOU
Patterson: KOKO LOVE WHAT?
Koko: LOVE THERE CHASE KNEE DO
Observer: The tree, she wants to play in it!
Patterson: No, the girl behind the tree!

Patterson's interpretation that Koko was indicating a wish to chase the girl behind the tree is not self-evident, to say the least.

It would be extremely useful to have real information on the abilities of gorillas to learn and use arbitrary symbolic gestures, and on the relationship between these abilities and other aspects of language and communication. Unfortunately, apart from a few data summaries produced in the first years of the project (when Koko's progress seemed parallel to that of Washoe or Nim), the Koko project has not provided such information.

Kanzi and Other Yerkes Studies

The studies we have been looking at so far attempted to teach nonhuman primates what the experimenters thought to be a natural human signed language. A somewhat different approach has characterized studies conducted at the Yerkes Regional Primate Center in Atlanta, Georgia. These were initially designed and carried out by Duane Rumbaugh and his colleagues, including his wife Sue Savage-Rumbaugh, who has become the principal scientist identified with this work.

What set these projects apart was that they did not attempt to teach ASL or any other naturally occurring language, but rather employed a completely artificial symbol system. It was based on associations between arbitrary graphic designs called *lexigrams*, presented on a keyboard connected to a computer, and meanings. Instead of producing a series of manual signing gestures, the experimental animal was expected to press the keys corresponding to what he (presumably) meant.

Prior to the lexigram studies, the general approach of devising an artificial system was tried out in David Premack's work with a chimpanzee. Sarah was trained to manipulate arbitrarily shaped and colored plastic chips on a magnetic board. Her impressive achievements included apparently learning the reference assigned by her human trainers to these chips, and developing categories of meaning. The relevance to studies of language has been widely acknowledged to be quite limited, however, and I will not treat it in detail. Its primary importance to our story is the way in which Sarah's plastic chips paved the way for later work with overtly artificial systems.

Duane Rumbaugh worked with Lana, who was the first chimpanzee taught "Yerkish," the keyboard-based language of lexigrams. Lana's training was intended in part to see whether she could learn a limited syntax. Some sequences of lexigrams were "grammatical" and others were not. Lana was supposed to produce expressions in this language to get rewards. She did achieve some success and, even more than Sarah, demonstrated skills in the domain of symbolic (and numeric) representation and reasoning.

A host of limitations on both the "language" and Lana's performance makes it difficult to draw serious conclusions about her linguistic abilities. The experimenters themselves considered that Lana had shown at least some syntactic ability, but even the most charitable interpretation of her utterances would not go beyond structure attributable to a very limited word-chain model. Rumbaugh and his colleagues have acknowledged that the Lana project was useful largely for what it taught them about research methodology.

A somewhat more significant experiment was then conducted using two chimpanzees, Sherman and Austin, who were trained by Sue Savage-Rumbaugh to use the lexigram keyboards. At first they learned to request things from each other, and later to name objects, though they seemed to have a lot of trouble transferring what they learned on one of these tasks to the other. Identifying a banana with a lexigram did not transfer directly to asking for a banana (with the same lexigram), for instance.

After a number of years of training, Sherman and Austin could do several things of interest, in addition to the appealing (though less cognitively significant) trick of using their keyboards to cooperate in obtaining rewards under complex circumstances. They could learn new lexigrams from observation alone, then use these lexigrams in new contexts. Further, they could use lexigrams to attribute properties (including color) to an object presented only through another lexigram. Thus, they could "say" that a banana is yellow without having to see an actual banana at the time. They could also classify lexigrams into one or the other of two groups depending on whether the referent was a food or a tool, strongly suggesting that the lexigrams had genuine meaning for the apes.

These results, certainly intriguing, were not particularly revealing about the presumed ability of chimpanzees (or other primates) to learn a real language. The constructed nature of Yerkish allowed the experimenters to avoid some problems presented by real (spoken or signed) languages, but the amount of structure present in the system is limited and certainly far from that in any real human language.

The research that stands apart from all of the other work with apes began when Savage-Rumbaugh began to work with Matata, a bonobo. Bonobos were long considered to be a smaller, "pygmy" form of chimpanzee, but primatologists have come to appreciate that they are actually a different species. Extremely rare in nature, they are lively and intelligent, and have a somewhat elaborate social organization in which males and females share food and child-raising responsibilities, engage in sex for social and not purely reproductive reasons, and display other traits rather atypical of their fellow nonhuman primates.

Matata was to be trained to use the lexigram keyboard like Sherman and Austin, but she turned out to be rather a poor student. Many long training sessions, with experimenters pressing lexigram keys on a keyboard connected to a computer (which responded by lighting up the key and also producing the spoken English word) and indicating the intended referent, seemed to get nowhere. Matata was evidently too old to learn this particular new trick.

Then something remarkable happened. Matata's infant son, Kanzi, was present during these training sessions, since he was too young to be separated from her (although he was considered more of a nuisance and a distraction than an experimental subject). When Kanzi was about 2½ years old, however, the unsuccessful Matata was removed to another facility for breeding. Suddenly Kanzi emerged from her shadow, showing that although he had had no explicit training at all, he had nonetheless succeeded as his mother had not. He had obviously learned how to use the lexigram keyboard in a systematic way. For instance, he would make the natural bonobo hand-clapping gesture to provoke chasing, and then immediately hit the CHASE lexigram on the keyboard.

From that point on, the focus of the work was on the abilities Kanzi had developed without direct instruction. His subsequent training did not consist of formal keyboard drills, with food and other treats as rewards for successful performance. Instead, the keyboard was carried around and the trainers would press lexigrams as they spoke in English about what they and the animals were doing. For instance, while tickling Kanzi, the teacher said "Liz is tickling Kanzi" and pressed the keyboard keys LIZ TICKLE KANZI. Kanzi himself could use the keyboard freely, which he did to express objects he wanted, places he wanted to go, and what he wanted to do. More structured interactions took place, as when Kanzi was specifically asked to "Show me the tomato lexigram" or to press a key in response to "What is this called?"

By the time he was about 4 years old, Kanzi had roughly forty-four lexigrams in his production vocabulary (according to a criterion that required consistent, spontaneous, and appropriate use), together with recognition of the corresponding spoken English words. He performed almost perfectly on double-blind tests that required him to match pictures, lexigrams, and spoken words. He also used his lexigrams in ways that showed clear extension from an initial specific reference to a more generalized idea. Thus, COKE came to be used for all dark liquids and BREAD for all kinds of bread (including taco shells). Certainly, further questions can be (and have been) asked about just what the lexigrams represent for Kanzi. Nearly all of the ones on which he can be tested for comprehension involve objects, not actions, so the richness of his internal representation of meaning is difficult to assess. Nevertheless, the lexigrams definitely appear to have a symbolic value.

Kanzi is reported to have used his lexigrams not just when interacting with an experimenter, but also when alone. He would take the keyboard away and press keys in private. He might press PINE-NEEDLE and then put pine needles on the key, press ROCK and put little rocks on the key, press HIDE and then cover himself (or the keyboard) with blankets. If a human attempted to interact with him while he was doing this, he would stop immediately. As a result, no systematic data exist on his private keyboard activities. We have anecdotes that are enormously suggestive, but no information about the possibility that he may have pressed the keyboard by himself many more times in random or otherwise unintelligible ways. The same can be said about the reports that Washoe and other chimpanzee subjects from earlier experiments made signs in private while looking through magazines and books of pictures. It certainly looks as if these animals are "talking" to themselves, but we need much more evidence to understand exactly what is going on.

Kanzi's Control of Syntax

Kanzi surely learned a collection of "words" in the sense of associations among an arbitrary shape (the abstract lexigram pattern), an arbitrary sound (the spoken English equivalent), and a meaning of some sort, and he can use these symbolically, independent of specific exemplars or other contextual conditions. Over the years, his vocabulary has continued to expand. His keyboard now contains 256 lexigrams, and his recognition vocabulary for spoken English includes many more words.

What can we say about Kanzi's potential syntactic ability? A major difficulty is that we need to assess two different and incommensurate systems, those of production and of recognition. Kanzi's production centers on the keyboard, and he understands a great many things in spoken English. He cannot, of course, produce English words, although he is reported to vocalize sometimes in ways that suggest an attempt to form spoken words. Let us look at each of these systems in turn for evidence of syntactic understanding.

When Kanzi uses his keyboard, he does not produce enough multilexigram sequences to permit true analysis of their structure. This is not to say that he does not produce complex utterances, however. In addition to his keyed lexigrams, he uses a number of natural, highly iconic gestures with meanings such as "come," "go," "chase." He also employs pointing gestures to designate persons, and he frequently combines a lexigram with a gesture to make a complex utterance. We might be able to analyze those combinations to see what emerges in terms of potential rules of grammar.

When we do so, we find some reliable tendencies, such as the orders action-agent, goal-action, and object-agent. These are somewhat unusual, for they certainly are not the orders that occur in Kanzi's input. English has agents preceding actions, not the other way around, and so on. In any event, a semantic analysis of these orderings is beside the point, because virtually all Kanzi's complex utterances of this type conform to a single overarching rule: lexigram first, then gesture. This principle of combination is intriguing, based as it is on the modality rather than the content of the symbolic expression, but it does not provide any support for syntax.

The principal evidence that has been cited for Kanzi as a syntactic animal comes not from his production, but from his comprehension of spoken English. An extensive study explored Kanzi's understanding in relation to that of a human child (Alia, the daughter of one of his trainers) at a similar stage of language development—at least in terms of vocabulary and MLU. A complete presentation and assessment of this study (and subsequent work on this aspect of Kanzi's abilities) requires far more space than we can devote to it here. One great advantage of the studies of Kanzi in general is that many of the relevant data have been made generally available, and those who are interested can explore the facts and come to their own conclusions.

Both Kanzi and Alia showed considerable ability to respond appropriately to requests like *Put the ball on the pine needles*, *Put the ice water in the potty*, *Give the lighter to Rose*, and *Take the snake outdoors*. Many of the actions requested (squeezing hot dogs, washing the TV, and the like) were entirely novel, so the subjects could not get along by simply doing what one normally does with the object named.

The range of possibilities correctly responded to by both Kanzi and Alia was sufficient to demonstrate that each of them was able to form a conceptual representation of an action involving one, two, or more roles (participants and/or locations) and then connect information in the utterance with those roles. This is the sort of representation of meaning that linguists refer to as a "thematic" description, with the individual participants associated with distinct "theta roles." It seems likely that many animals have internal representations of complex concepts with this character, but Kanzi is the first nonhuman in whom we have evidence for an ability to link the various parts of such a representation with parts of a communicative expression.

We can also see that the connections Kanzi makes between parts of what he hears and parts of a complex, thematically structured concept respond to some extent to the form of the utterance. He can satisfactorily distinguish between *Make the doggie bite the snake* and *Make the snake bite the doggie.* At a minimum, he must be sensitive to regularities in the order of words; he did not simply interpret the content words of a sentence in their most familiar way, or in some consistent, invariant way.

These facts provide evidence for something like a word-chain model, which has regularities in terms of what follows what (for instance, agents precede actions and objects follow them). This is a totally unprecedented result in the literature on animal cognition, but it does not in itself argue that Kanzi represents sentences in terms of the kind of structure we know to characterize human understanding of language. Much of what we see might not rely on any particular structure, but rather result from a sort of "substitution in frames" procedure. That is, perhaps Kanzi has learned that certain complex utterances have places in them where there is room for one of a small set of different possibilities. Such an analysis would not require any appreciation of hierarchical organization, constituent structure, or the like. The range of patterns on which Kanzi has been tested is limited, but very little in the way of structural knowledge seems to be required.

In fact, on those sentences whose interpretation depended on information provided by grammatical words, such as prepositions or conjunctions, Kanzi's performance was quite poor. Distinctions such as that between putting something *in, on,* or *next to* something else appear not to have been made. Sentences with *and* (whether conjoining nouns, as in *give the peas and the sweet potatoes to Kelly,* or sentences, as in *go to the refrigerator and get the banana*) frequently resulted in mistakes of a kind that suggest such words simply went uninterpreted.

One class of sentences on which Kanzi did well supposedly showed his ability to understand the structure of relative clause constructions: *Go get the carrot that's in the microwave.* But it does not follow from his ability to respond appropriately to this request that he has understood it on the basis of a hierarchical structure with an embedded relative clause. If we attend only to the content words here (*go get, carrot, microwave*) and try to fit them into a semantic schema, *carrot* obviously has to be the object of getting, but *microwave* has no role to play in that action and can only be interpreted as a property of the carrot (its location). A coherent interpretation requires an appreciation of meanings and their thematic structure, but not of specifically grammatical organization.

Actions and objects (as represented by concrete verbs and nouns) correspond to things in the world, and they can constitute the meanings of symbols for Kanzi. Grammatical markers, however, get their importance not by referring to something in the world, but by governing the way *linguistic* objects are organized. Kanzi has a method for associating the referential symbols he knows with parts of complex concepts in his mind when he hears them. This method does not involve genuinely grammatical structure, so "words" that have significance solely in grammatical terms can only be ignored.

It may seem that I have gone to great lengths to avoid the conclusion that Kanzi has a meaningful appreciation of the grammar of English, given that he can apparently understand many English sentences. It is certainly not my intent to underestimate the interest and importance of the abilities that Savage-Rumbaugh has demonstrated and carefully documented in Kanzi. But while the evidence available takes Kanzi far beyond the other animals whose cognitive and communicative abilities have been studied, it does not in fact show that he has acquired an understanding of the syntactic structure of a natural language. Without that, he cannot be said to have acquired language in its core sense.

Apes and Language

Having surveyed the evidence that is available from the attempts to teach apes a human language, we can now draw some conclusions. Apart from

Savage-Rumbaugh's ongoing work with Kanzi and other bonobos, it is unlikely that further projects of this sort will be undertaken in the near future, in part because of the perceived air of failure that surrounds the earlier efforts. That is unfortunate: while it seems evident that apes do not have the specialized cognitive faculty that would allow them to "learn language" in a complete way, the research has demonstrated abilities in these animals that had not previously been suspected, and about which it would be exciting to learn more. It may be that at least some of the limitations of the existing body of evidence are limitations of the experiments, and not necessarily of the subjects.

Some factors are obvious. No ape can learn to *speak* a language like English, because the anatomy of their vocal tracts is incapable of producing the relevant range of sounds. Some factors are less obvious, but probably true (and relatively uncontroversial). Apes reach a plateau as far as complexity of expression is concerned. No matter how extensive the training, no animal is going to produce long, complex sentences. If we want to know whether an ape can develop an ability to use a human language that is comparable to that of even a grade-school child, the answer is a definite no.

But we can ask a different question: Do the apes in these experiments show evidence of having learned something that has significant resemblance to human language—a system that has some properties human languages have, and naturally occurring systems of animal communication do not have? Let us enumerate the essential components of our knowledge of language, then look for evidence in the ape experiments that bears on the animals' achievements with respect to each element.

Our knowledge of language includes at least the following:

- *Lexicon:* a collection of *words*, in the sense of a set of arbitrary associations between external expressions (in sound or signs) and meanings.
- *Phonology:* a discrete combinatorial system that supports the combining of formative elements (sounds or the formational components of signs, including handshape, location, and the like), taken from a small basic set, into expressions that are linked to meaning as words.
- *Syntax:* another discrete combinatorial system, which licenses the combining of words into phrases, of phrases into larger phrases, and so on. This system derives its force from the fact that it is based on word classes, grammatical relations, and other properties. In particular, it is

recursive, so it accommodates an unlimited range of distinct sentences on the basis of a relatively small set of "known" words and rules.

In this listing I have more or less left out semantics, the principles by which the meaning of a complex expression is determined on the basis of the meanings of its parts and the manner of their combination. Unless a system includes complex syntactic structures, it makes little sense to explore the ways in which these might be assigned an interpretation. I have also left out principles like those that determine the interpretation of pronouns (see Chapter 3). These and other aspects of human knowledge rest on the foundation of syntactic structure, so the first aspect to explore is whether apes have a system with that essential structure in place. It does not make sense to ask whether they can learn how to interpret pronouns if they do not have knowledge of the kind of structure on which the working of that system rests.

Postponing the question of a lexicon for the moment, let us start with the matter of a phonological system. Do any of the animals we have discussed have a discrete combinatorial system at the base of their meaningful communicative expressions? In the case of lexigrams such as those employed by Kanzi (and before him, Lana, Sherman, and Austin), there is no question of any system. The lexigrams are carefully constructed, in fact, so as to consitute unanalyzable wholes. In the case of signs, we have seen that the apes get these structural matters wrong, and get them wrong in ways that suggest they do not grasp the notion of a specific set of formational elements.

For instance, the animals in these experiments show no awareness of the fact that in a language such as ASL certain handshapes are possible and others are not. When the apes make up novel signs, as they sometimes do, or distort the form of signs they are shown, there are no obvious constraints on the shape their hands adopt apart from those of physiological necessity. Recall that in ASL the difference between basic forms of pronouns (I, you, he/she/it) and possessive forms (my, your, his/her/its) is systematically a difference between a pointing and a flat handshape. While some of the apes have learned MY in relation to I, they show no appreciation of the generalization of that difference to YOUR, HIS, and the rest. In general, we find no evidence of any combinatory system underlying the expression system of any of the apes. Indeed, we will suggest in Chapter 11 that this absence may be related to the fact that their vocabularies seem to be limited to a few hundred signs at most—small in comparison with the lexicon of even a rather young child.

What about the special case of Kanzi, who clearly recognizes a variety of spoken words? I argued in Chapter 5 that speech recognition in people is based on a motor theory, and on a decomposition of the speaker's activity into abstract formational elements of motor control. Of course, the reason we make this kind of assumption about humans is in part because of the speed, efficiency, and flexibility with which we recognize an unbounded range of possible sound combinations. Because Kanzi does not have more than a few hundred words (on the most optimistic assessment) to distinguish, no such argument is valid.

Savage-Rumbaugh has argued that Kanzi has a "phoneme-based" system for recognizing words, an argument that I find extremely weak. What she did was present him with three choices for a spoken word: the correct choice, one that shared the beginning sound, and one that shared the final sound. Thus, *paper* might be the stimulus, and *paper, peaches, and clover* the possible responses. Kanzi did very well at choosing the original word correctly, but what does that prove? It just shows that he can discriminate among (holistic) acoustic patterns that overlap somewhat in physical form. There is no reason to presume that any analysis of the internal structure of the pattern is responsible, for none is necessary. Many animals actually can learn to discriminate members of a small closed inventory of human vocalizations—just as we can learn to discriminate theirs.

What about syntax? Do the animals in these studies develop a discrete combinatorial system? That would require that they combine elements, of course. Discrete elements. And that they combine them according to a system, one that is based on generalizations such as the fact that nouns behave in one way and verbs another; and that noun phrases have the same form regardless of whether they are used as subjects, objects, or in some other grammatical function.

We must distinguish the animals' production from their recognition ability, since the evidence is somewhat different in the two cases. In terms of production, the range of their sign combinations is rather limited. Furthermore, the predominance of repetition in longer sequences suggests something like the semantic soup view: at a given time many things are salient to the animal, who makes signs (or chooses lexigrams) that correspond to them—but individually, rather than as a complex internally structured whole.

What we need in order to establish a syntactic view of the animal's competence is a set of rule-governed regularities. What we get, however, is at best statistical regularities. Some ape language researchers argue that their animals behave in a way that corresponds to early stages of language acquisition in human children. However, the regularities in children's speech are categorical, not merely statistical tendencies.

An exception may be Kanzi's combinations, which seem to reflect the genuine rule that "lexigram comes before gesture." This is, however, a strange sort of rule, since it involves not two distinct grammatical categories, but two quite different modalities. Apart from this single odd example, the other regularities we find look more like word-position preferences (YOU before ME) than like structurally based regularities (subject-verb-object). The proposed objection that the lack of regular order in the animals' productions is related to the fact that ASL has free word order does not survive examination, since the apes did not have ASL as input and they did not produce the specific devices that ASL uses. The bottom line is that there is little or no evidence for any real combinatory structure in the productions of any of these animals.

On the perception side, by far the best evidence is the set of perceptual tests given to Kanzi. I suggested above that Kanzi's recognition system for English allows him to make connections between spoken words and particular roles in a semantic (or thematic) structure. Furthermore, the connections he makes are sensitive, to some degree, to word order. From these facts we conclude that he may have structure of the sort we should call a word-chain model. If confirmed in further research, this would be a remarkable fact; no other nonhuman animal has plausibly been shown to do better than semantic soup on the informal scale we have been using. It is still a long way from syntax of the sort found in human languages, however.

Much more would need to be shown before we could accept the claim that Kanzi (or any other animal) has a real appreciation of the syntactic form of sentences in a natural language. To say that is not to denigrate his remarkable achievements, or to cling to an outmoded exaggeration of human uniqueness. It is merely to require evidence commensurate with the capacity that is being attributed to him. Unfortunately, those who conduct these experiments are often unfamiliar with the real nature of syntax in human languages, and they tend to accept any sort of demonstrated combination of meaningful elements as "syntactic" enough to count as language-like. If one believes that syntax is simply a matter of putting words (or signs) together one after another, the burden of proof is not huge; but that is not what is at stake in claims for syntactic ability in nonhuman animals.

We must conclude that the parts of language that form discrete combinatory systems, including phonology and syntax, seem not to be accessible to the primates that have been the objects of investigation. I have ignored another combinatory system in natural language here, that of *morphology* or word formation. Words are commonly formed from other words according to patterns of modification that can be cumulated to produce very complex structures internal to a single word. We saw in Chapter 9 that ASL has a rather complex morphological system, and it would certainly be relevant to know whether such systematic relations among classes of words could be appreciated by a nonhuman subject. In the absence of phonology and syntax, it seems highly unlikely.

What about a lexicon? What evidence is there that apes can use a set of arbitrary signs in the kind of way speakers of human languages do, to refer to concepts, objects, and relations in the world? To establish this thesis, we need to show symbol use that meets at least the following conditions:

- *Noninstrumentality:* The symbols are genuinely used to *refer* to something, not simply as a means for carrying out some action or getting something.
- *Displacement:* The symbols can be used to refer to things that are not necessarily present in the environment when used.
- *Noniconicity:* The symbols are not direct representations of what they represent in the world.

The last two are perhaps obvious requirements for treating gestures or lexigrams as "words." To see the importance of noninstrumentality, imagine what happens when I go to the vending machine in the basement, insert money, and press the buttons A-0-9 in sequence to receive a package of M&Ms. This is one possible interpretation of the situation in which a chimpanzee presses a prescribed sequence of lexigrams on a keyboard and receives a reward. Both of us press a sequence of buttons, in my case labeled A and then 0 and 9, for the chimpanzee having abstract symbols. The chimpanzee has learned the sequence from many trials, gradually built up from a single symbol, while I have the advantage of being able to read A-0-9 on the slot with the M&Ms. Can I interpret my gestures "Insert money," "Press A," and so on, or the corresponding button presses of the ape, as "utterances" like "Please machine give me M&Ms!"?

In both cases, interpretation of the sequence of buttons pressed as essentially equivalent to an English sentence ("Please machine give me M&M's!") is wishful thinking at best. What is going on need in no way involve the essential properties of a language. It is just a routine we go through to get M&Ms (which both the chimpanzees and I like, and are willing to go to some lengths to obtain). To the extent that an ape's utterances all have this character—and by and large, those of the signing chimpanzees do—they do not represent what we do with language.

Most of the apes' utterances are instrumental: ways to get food or treats, including being taken places or other enjoyable experiences. Even Kanzi rarely seems to comment on the passing scene or to ask questions out of curiosity. In virtually all instances, his utterances are intended to get something. The major exception seems to lie in the reports by Savage-Rumbaugh or the Gardners of times when an animal sits quietly by himself paging through picture books or magazines, and sometimes makes signs or presses keys that correspond to what he sees. To the extent that this behavior can be seriously documented, it constitutes genuinely noninstrumental use of signing.

Perhaps, indeed, the fact that most of the signing observed in language-trained apes is unambiguously directed at obtaining rewards says more about the nature of the relationship between the animals and the humans who study them than it does about cognitive or language abilities. From the animals' point of view, the humans may be around mostly to provide food and fun, and the reason the apes learn to make these gestures is to ensure their supply of these benefits. They may well be able to use their signs in other ways (and there is limited evidence available to suggest that that is the case), but most of what human experimenters see illustrates only instrumental uses.

As for noniconicity, it is not seriously in doubt. Kanzi's (or Sherman and Austin's) lexigrams, for example, are wholly noniconic. If we accept that the apes have a sense that the lexigram is a sign for something, it is obviously noniconic. And in the sign experiments, while many of the gestures the animals use represent their referent directly (pointing gestures, touching parts of the body that are to be attended to), and still others are naturally

occurring (probably innate), many others are likely to be learned arbitrary associations. The learned part is presumably important: our vocabulary has the open-ended quality it does because we can learn new words and are not limited to a fixed, innate set. Some of the chimpanzees' signs are apparently ones that occur in nature and those are presumably innate. If those were *all* the animal had, they would not constitute much of a vocabulary — but they are not.

On balance, there does seem to be considerable evidence that the animals in these experiments have learned a set of arbitrary symbolic expressions, even if their primary use for them is to get what they want. It is still a rather remarkable ability, apparently not displayed in nature. I shall return to this point in the closing chapter of this book.

Alex the Parrot

"Stubbins is anxious to learn animal language," said the Doctor. "I was just telling him about you and the lessons you gave me when Jip ran up and told us you had arrived."

"Well," said the parrot, turning to me, "I may have started the Doctor learning but I never could have done even that if he hadn't first taught me to understand what *I* was saying when I spoke English. You see, many parrots can talk like a person, but very few of them understand what they are saying. They just say it because — well, because they fancy it is smart, or because they know they will get crackers given them."

— The Voyages of Doctor Dolittle

One of the more fascinating and (to my mind) significant animal "language" studies deviates markedly from the ape language studies we have focused on in this chapter. Since the late 1970s, Irene Pepperberg has been working with an African grey parrot named Alex. Her research is reported in detail in her book *The Alex Studies*.

The activity of most "talking" parrots, mynah birds, and others is relatively uninteresting from the point of view of language. These birds can learn to produce some noises that humans hear as sentences, but whatever meaning these productions may have for the bird has nothing to do with what the sentences mean to us. Indeed, the acoustics of this bird "speech" differs interestingly from normal speech, though there are also similarities. Given the differences in human and avian anatomy, the mechanisms of production are significantly different as well, although unlike most other animals, a parrot does manipulate the shape of its vocal tract in forming different sounds. Arguably, despite the variations of these acoustic signals from actual speech, they nonetheless have the acoustic characteristics necessary to engage the special speech mode of auditory perception discussed in Chapter 5, and thus to be interpreted by humans as speech.

Alex has apparently learned a substantial vocabulary of color words, numbers, names for objects, shapes, and the like. More to the point, he can deploy these words so as to answer questions, ask for objects, and say what he wants. He has probably not acquired anything much in the way of syntax (Pepperberg explicitly avoids the claim that Alex "has language"), but the obvious potential problems with this research (such as the possibility of a Clever Hans effect) have been ruled out. Alex seems to be the genuine article, suggesting that in an animal capable of producing speech-like sound with some fluency, a surprising amount of language-like behavior can be elicited.

Recall that the ape sign language projects were originally started on the basis of the premise that apes had enough cognitive capacity to learn language, but could not deal with the articulation of speech. The opposite would seems to be true for a parrot. These birds produce sound in somewhat different ways from humans, but they can imitate a wide range of sounds in a readily recognizable way.

Pepperberg was working on her doctorate in chemical physics at Harvard University in the 1970s when she heard (via a Nova program) about the signing ape projects, and decided that they sounded like more fun than what she was doing. She took courses in avian biology and related subjects, and after getting her degree and moving to Purdue University, she bought Alex in a Chicago pet shop. The project started at Purdue, moved to Northwestern University in 1984, and then to the University of Arizona in 1991. In 1999 she and Alex moved to the Media Lab at the Massachusetts Institute of Technology, where in addition to language, they worked on a Web browser for parrots. As of this writing their research is continuing at Brandeis University.

A major aspect of this project is the training model Pepperberg originated. Building on earlier work by the German ethologist Dietmar Todt, she developed a competitive ("model-rival," or "M/R") technique of interaction, which has proved to be her key to success in this endeavor. On this approach, the researcher and an assistant interact with each other in the parrot's presence, an activity that seems to be highly motivating. The parrot wants to play too, and wants to learn how to get the objects the humans have, as well as generally seeking their attention and approval. Through this training regime, Alex has learned the names for a number of objects, which he produces appropriately. Considerably more interestingly from a cognitive point of view, he has learned names for a number of colors, shapes (expressed in terms of number of corners: "four [corner]" for "square"), materials, the numbers through six, "none, no" and much more.

What can Alex do? He can label objects ("key," "nut," and so on). When he does this correctly, he usually gets the object named, which he may eat or simply chew on (parrots are fond of chewing or gnawing on things). He can ask for what he wants, when it is not present ("want nut"). He can identify the shape (2, 3, 4, 5, 6-corner), material ("wood," "paper," "cork"), and color of an object. Presented with an array of things on a tray, he can give the number of objects in the set. More dramatically, he can give the number of objects that meet some criteria ("How many four-corner wood?") out of a larger set. When appropriate, he can identify the answer as none ("No"). He can classify colors, shapes, materials, and quantities (numbers) together. Perhaps his ultimate tour de force is the following: presented with a diverse collection, he can identify the dimension with respect to which the objects are similar or different ("color," "matter," and the like).

How should we characterize the communication system Alex has acquired? He has an inventory of individually meaningful words, rather than a set of holistically interpreted utterances. He often makes errors that consist in leaving out a word ("four" is a common error for "four corner" in

"What shape?" questions), which suggests that the words have a sense by themselves and not just in a specific context. He clearly has a system in which these words are combined to form larger wholes. We have no reason to believe in anything like internal constituent structure, but his internal grammar must have (at least) the properties of a word chain as far as receptive capacity for syntax is concerned. This trait is all the more meaningful in light of the absence of evidence for anything so complex in the behavior of most other animals.

What should we say about the nature of Alex's "words"? They are certainly noniconic (as opposed to many of the gestures seen in the signing chimpanzees), since the acoustic products of his (and the experimenters') vocalizations have no intrinsic connections with what they refer to. *Do* they "refer" to something? Evidence in favor of that interpretation is that when he asks for a nut and the experimenters give him something else, he can say "No. Want nut."

Are Alex's utterances instrumental, in the sense that he produces them as a way to obtain a reward? Largely so. Pepperberg stresses that when his answer to a question is correct, he gets what he named: that is, his rewards are intrinsic, not extrinsic. When the object named is one that does not really interest him and he answers correctly, his reward is the right to ask for something else. This procedure makes it a bit more circuitous to interpret his utterances as directly instrumental, in the sense of producing a direct reward. And Alex does vocalize when he is alone, even engaging in what seems to be verbal play with the sound patterns he uses in interaction with the experimenters.

The most interesting results to date as far as cognition is concerned involve Alex's ability to establish higher-level categories such as shape, color, and number. Work currently under way is attempting to teach him to use visually presented arbitrary symbols (such as Arabic numerals for numbers) for the categories he already knows verbally. Essentially, his trainers are trying to teach him to read. Other parrots are now involved in the same training, and Alex is serving as one of the tutors.

Pepperberg has no illusions that Alex is learning English. Rather, she is interested in exploring the possibilities of using English words as a code for "interspecies communication" in order to learn about concept formation and other aspects of the mental life of an animal. That is, she is interested in exploring the parrot's cognitive abilities, and in that endeavor, (some aspects of) language can serve as a tool, rather than necessarily as the object of inquiry.

This seems to me the best kind of language-related research to pursue with animals. There is no reason to believe that human language per se is accessible to other animals. It is always possible that we will learn differently at some point, and novel training methods could show the way toward some such result, but basically animals do not learn language in anything like the sense we do. On the other hand, we *can* use their communicative abilities to ascertain more about animal cognition.

Alex is truly a remarkable bird. Yet when we compare the abilities he has shown with those that have been demonstrated in language-trained chimpanzees, the contrast is at least superficially dramatic. It is hard to believe that the overall cognitive skills of parrots are more sophisticated than those of chimpanzees, so we can only anticipate that different approaches to our evolutionarily closest kin will eventually lead to much more exciting insights into the primate mind. The same conclusion is supported in a limited way by the finding that Nim's signing was somewhat more spontaneous and interesting in a conversational setting than in the setting of explicit training. It would seem, perhaps, that we need to abandon the approach that sees "learning language" in a human sense as the only worthwhile goal, and use the communicative abilities that animals can acquire as a window into their cognitive processes more generally.

11

Language, Biology, and Evolution

"Ah!" said the Doctor. "I had guessed it would be Noah, but for what reason, Mudface, was he made zoo-keeper?"

"Because," the turtle answered, "he spoke animal languages. He was the first and only man, besides yourself, John Dolittle, who could understand animal talk. Having lived six hundred years, you see, he had plenty of time to learn."

"Of course, of course," said the Doctor. "And how wonderfully well he must have learned to speak in that time!"

"No, there you're mistaken, Doctor. He could make himself understood, it is true. But he could not chat in turtle talk, for example, half as well as you can—nor even as well as Tommy here. And as for writing

down things, he never learned. He was not quite as wise and clever as he
was cracked up to be."

—Doctor Dolittle and the Secret Lake

To paraphrase Art Linkletter, animals do the darnedest things. When they
communicate with one another, some of what they do is quite outside our
own capabilities. Even if we ignore exotica such as ultrasound in bats, elec-
tric fish, and the like, for which we lack the relevant sense organs, consider
the use of infrasound by elephants for long-distance communication, the
underwater songs of whales, or the olfactory communication of lemurs. We
have also witnessed some of the remarkable (and perhaps unsuspected)
complexity of human language, though. For those of us who are human
language users, our linguistic ability seems absolutely straightforward, but
from the perspective of another species (if it were possible to adopt such
a point of view) it would surely be as mysterious and inaccessible as, say,
electroreception in fish seems to us.

For many readers, the argument should end right here. We know that
various species, by virtue of their own individual biologically determined
nature, have communication systems that are suited to their distinctive life-
styles and ecological problems. *Homo sapiens* is one such species, and its
members use what we call natural language (and not electroreception or
ultrasound) as a fact of our nature.

For others, though, the notion that communication systems (an aspect
of behavior) are as much a part of an organism's biological identity as, say,
the relative length and position of its thumb, is unsatisfying. An animal's
body is structured by processes that have accumulated over evolutionary
time, but its behavior, this line goes, is constrained only by very general cognitive faculties that could in principle be deployed in ways quite different
from those we happen to observe.

As a result of this discomfort, much research on animal communication
has been driven by a need to show that there is nothing uniquely human
about human natural language. Other animals, especially those closest to
us, could perfectly well control such a system if (a) relatively superficial
constraints, such as the lack of a human-like vocal tract, were removed, and
(b) they were given the basic idea of a human language to work with.

The structure of natural language is just another skill to be developed
by general learning mechanisms and applied in accordance with the organ-

ism's overall cognitive capacity, this theory goes. True, it is unlikely that these general mechanisms as they exist in, say, squirrels are up to the demands of learning to speak French, but in higher primates at the very least, language ought to be accessible in its fundamentals. The claim that language is possessed only by humans, like so many other claims about human uniqueness that have fallen by the wayside over the years, is another example of a misplaced superiority complex on the part of human scientists and philosophers.

An a priori argument due to Chomsky might give us pause here. He reasons that animals *could not* have the cognitive capacity to learn a language (in the sense of a human natural language), in that they never display this capacity in nature. To say that an animal *could* manifest an ability as evolutionarily advantageous as language, but simply has never done so, is as ridiculous as saying that somewhere an island exists on which birds are perfectly capable of flight, but have not yet thought to do so and need human instruction to induce them to fly.

Although some (especially in the ape language research community) have rejected this argument out of hand, it seems persuasive to me, and what we have seen in this book so far is fully consistent with its premise. No animal has been shown to control, either naturally or under conditions of laboratory research, a system with the central properties of human language. We conclude that there is a simple reason for this: Language as we know it is a uniquely human capacity, determined by our biological nature, just as the ability to detect prey on the basis of radiated heat is a biological property of (some) snakes.

In this concluding chapter, I make several distinct points with respect to this important conclusion. First, I raise some of the many questions about how the human language faculty arose in the first place. The problem of the evolutionary origins of language has a complex and troubled past, and is anything but straightforward in the present. Nevertheless, if we want to maintain that language as we know it is a biologically determined property of our species, it is essential to sketch how that might have come to be the case.

I suggest a further consequence of the logic of Chomsky's argument: if animals truly were capable of deploying a system like that of human language, surely some of them would have done so by now. But what does "like that of human language" really mean? When we look at what gives human language its expressive power, we come to the (somewhat nontraditional) conclusion that it is syntax, not symbolism, that is most important. And that is precisely what animals do not seem capable of.

Where Did Language Come From?

Throughout this book I have been promoting the idea that the human language-faculty is a part of our biological constitution, a specific faculty that is precisely tuned to the nature of language and that no other animal has. So the logical question is, where did it come from?

Although it is an obvious question to ask, how to go about answering it is much less obvious. Language does not leave any traces in the fossil record or elsewhere of the sort that normally guide accounts of the course of evolution. Acts of speaking (or signing) do not leave anything behind that we could dig up thousands of years later. Even the apparatus we use for speaking is largely elusive; it is almost entirely composed of soft tissue that leaves no long-term residue either.

For a long time these problems did not stop people from speculating about how language *might* have arisen. This was actually a rather popular topic among Romantic philosophers. With the rise of a more scientific approach to language in the nineteenth century, discussions of language that were not based in empirical fact came to be discouraged. Famously, the Linguistic Society of Paris, in its constitution of 1866, said that papers presented to the society could deal with any topic concerning language except two. These were the origin of language and the construction of international auxiliary languages like Esperanto or Volapük. What these topics have in common is that (at least as far as people could judge in 1866) both are matters of opinion and speculation, not science.

Were the founders of the society right in barring language origins? Or can we say anything serious about where the language faculty came from? If we believe in literal biblical creationism, there is no problem. But if we are not satisfied with that answer, we have to imagine that it arose during a much longer course of human evolution.

Saying that the language ability arose *during the course of evolution* is not exactly the same as saying that human language arose *through evolutionary mechanisms*, and there are in fact two divergent views on the matter. One school, with which Chomsky has become associated, holds that language is so unusual, and so unconnected with anything else in the animal world, that it could not have arisen gradually through evolutionary pressures. Instead, some other combination of changes in, say, brain size and complexity must have yielded language as an accidental by-product in a sudden, discontinuous way. This position is not popular, because without a proposal for just what those factors might be, and how language could arise as a consequence of something like overall neural complexity, the question is reduced to pure mystery.

Unfortunately, that might well be the right answer. Chomsky has suggested on several occasions that within the range of questions we might ask about the world, there is a difference between *problems* and *mysteries*. Problems are questions that we can ask, and for which we can seek evidence that might lead us to answers. Mysteries are questions that we can formulate perfectly well, but to which there may not *be* any answers. Either no possible evidence exists that could bear on the mystery, or our own cognitive limitations would prevent us from making sense of the answer — something like Douglas Adams' "42" as the answer to "life, the universe, and everything."

The alternative, associated with Steven Pinker, Ray Jackendoff, and others, is to say that language *must* have arisen through the normal mechanisms of Darwinian evolution — or at least that that is the only possible scientific approach. Within the last decade or so, this area of interest has become increasingly popular. Some of the writing is not particularly serious, because it is not informed by much knowledge of what the human language faculty really is like; but with linguists participating more and more in the discussion, the level of sophistication has risen significantly.

If we take the origin of language to be a matter subject to scientific investigation, we can probably assume that it corresponds to a unique set of events, rather than a number of parallel but independent inventions. Otherwise we would probably find at least some trace of differences among languages today that reflected that independence. Learned song arose independently in parrots, hummingbirds, and oscine songbirds to fulfill similar functions, but the song systems differ in their structure and underlying neurobiology. Eyes arose independently in mammals, insects, and the octopus because it is advantageous to be able to extract information from the visual environment, but the three solutions are very different in their basic structure. Similarly, the advantage of being able to communicate in a flexible and open-ended way might create pressure to develop language in creatures with a generally high level of cognitive complexity, but we have no evidence that more than one kind of language exists. On the contrary, the same principles of Universal Grammar seem to obtain in all human languages.

So when did the invention of human language take place? It must have been prior to the separation of any particular group from the rest of human-kind, if the modern descendants of that group can be shown to have a language (as all human populations do). The oldest such separation that we know of is the settlement of Australia, some fifty thousand years ago. That event gave rise to a population that was essentially isolated from the rest of the world, because the land bridge over which humans originally entered Australia was submerged shortly afterward. Australian aboriginal languages are rich, varied, and old, but they are undeniably languages and governed by the same language faculty as others. Therefore, language must have existed in its essential aspects before this separation.

We cannot declare that the original settlers of Australia did not have language when they went there and either (a) later visitors taught them to speak or (b) *they* were the real source and later on taught others to speak. Such a statement is impossible because we are talking about the origin of the language *faculty*, not about particular acts or ways of speaking. If the original inhabitants of Australia did not have that faculty when they separated from the rest of the species, they would not have been able to benefit from instruction at a later time—any more than nonhuman primates (who lack our language faculty) can.

The Origins of Speech

One crucial component of the emergence of human language is the special-ization of our vocal tracts for use in speaking. As I argued in Chapter 5, human speech is only possible with a human-like vocal tract; without it an animal is unable to produce most of the sounds that occur in human languages. Although the principal organs that we use for this purpose are shared with a wide range of other animals, significant differences also occur. Generally cited in this connection is a larynx positioned low in the throat, making a large, L-shaped pharyngeal cavity within which the tongue can move. The usual wisdom is that this property is uniquely human and distin-guishes us from other animals. A second important adaptation is the capa-bility of completely closing off the nasal cavity by means of the velum, so as to produce nonnasal sounds.

Given these differences, it makes sense to ask how they arose. Two overall scenarios seem to be possible, each of which raises a "chicken and egg" problem. Either the organs of speech took on their modern form in advance of the emergence of language and thereby facilitated that development, or language emerged first and the vocal tract adjusted itself to the needs of the new system. In either case we need to do some explaining.

The notion that an essentially modern vocal tract came first, as a convenient preadaptation for speech, is championed by Andrew Carstairs-McCarthy. He suggests that the position of our larynx and attendant modifications of the basic primate anatomy were due to the development of erect, bipedal posture, and that this "favored a reorientation of the head in relation to the spine and hence a shortening of the base of the skull, so that the larynx had to be squeezed downward into the neck." Since humans espoused bipedalism more than three million years ago, our ancestors would have had plenty of time to seize on such a change to develop articulate speech as the principal medium for language.

This theory has real problems, however. Tecumseh Fitch points out that kangaroos and gibbons also have essentially bipedal posture, with a head oriented to the spine in the same way as ours, but they do not have a lowered larynx. And although Pekinese dogs and Persian cats have been bred to have shortened snouts, this adaptation apparently has not resulted in corresponding pressure for them to develop a lowered larynx.

The extent of human uniqueness in this area turns out to have been seriously exaggerated in the literature. While humans are unusual in having a lowered larynx *at rest*, many other animals lower their larynx in the production of at least some vocalizations. Fitch studied dogs, goats, pigs, and cotton-topped tamarin monkeys, and found that when they *bark*, or *baah*, or otherwise vocalize, all of these animals come closer to human vocal tract configurations than is usually thought. They actively pull their larynx down into the throat to extend the length of the vocal tract.

Why do they do that? It seems that vocal tract length is usually a fairly accurate clue to overall size, and this dimension has an effect on the acoustics of the animal's vocalizations. The formants (or natural resonances) of the vocal tract will be systematically lower if the vocal tract is longer; thus, an animal that acts in such a way as to lengthen its vocal tract will sound bigger (and presumably more threatening). Lions, tigers, and other "big cats" have a particularly flexible structure that would, in principle, allow the larynx to be lowered a great deal. Fitch speculates that this is precisely what happens when they roar. Obviously, it is difficult to conduct MRI studies of roaring lions in order to confirm this hypothesis.

Further, while the nasal passages are coupled with the vocal tract in most animals at rest, when they vocalize they change the orientation of the head in a way that makes it possible to close off the nasal cavity with the velum. The nasal cavity absorbs a lot of sound, so shutting it off allows the vocalization to be much louder. The initial conclusion that animals could only produce nasal sounds thus turns out to be incorrect.

Finally, there are two species of deer in which the larynx of postpubertal males is permanently lowered, in a way quite similar to humans. In these species (red deer and fallow deer), the acoustic consequences of a long vocal tract have a demonstrable effect in intimidating other males, and in getting the female deer excited. Koala bears have permanently lowered larynxes, too. These animals are not, in fact, the teddy bears they may seem from far away. Male koalas use loud bellowing calls to advertise their presence and intimidate potential rivals, an activity for which the appearance of greater size provides an obvious advantage. (Apparently the only vocalization produced by female koalas is a harsh, wailing distress call when they are harassed by adult males.)

A lowered larynx is not unique to humans after all. It merely looks that way if we base our conclusions on the study of dead animals, permanently at rest and with their tissues essentially paralyzed in formaldehyde. Still, at least among primates, humans are the only creatures with a *permanently* descended larynx—others can pull the larynx down, but not all the time. So instead of assuming that the lowered larynx is a human innovation, resulting from something else (such as erect bipedal posture), we can postulate a more plausible theory: early hominids had the capacity to lower their larynx when vocalizing, and as speech increasingly became a feature of human life, it became more efficient, faster, and more accurate to have the larynx adopt that position all the time.

On this theory, it is the development of language and speech that affected the shape of the human vocal tract, not the other way around. This premise does not need to deny the capacity of speech to Neanderthals, as Philip Lieberman proposed some years ago on the basis of a study of fossils that suggested a high position for the larynx. Physical anthropologists have largely rejected Lieberman's interpretation, but if Neanderthals did have a high larynx at rest, they might still have been able to lower it when vocalizing and thereby come close to modern human phonetic capacities. They may or may not have had language, but we cannot resolve the matter by saying that they were incapable of speech.

The Rise of Modern Language

The best guesses in the existing literature suggest that language in more or less the modern human sense arose in our species within the last two hundred thousand years or so. As a result, comparative evidence is not going to help us much with its origins, since we are separated from chimpanzees, for instance, by roughly six million years. When we look at our fellow primates, we do find some of the bits and pieces of language: communicative use of gesture and vocalization, a certain amount of structured meaning, an appreciation of relatively complex meanings, limited counting ability. In laboratory experiments with some primates, we can induce the use of arbitrary abstract symbols, a capacity that they apparently never utilize in nature. But we do not find even rudimentary forms of language itself.

We do learn just how much innovation was needed in order to get to modern language. Humans, for instance, are the only primates who can manage vocal imitation (though some nonprimates, like birds and also dolphins, are quite good at it). This is a skill that surely plays a major role in word learning. We also find that nonhuman primates are poor at attributing significance to the kinds of meaningful gesture we make, such as pointing—although dogs (but not their close relatives, wolves) can do this rather well.

Another candidate for a difference between us and our fellow primates is that in lemurs, monkeys, and apes, vocalization is likely to be under the control of involuntary, subcortical structures in the brain. Some of these animals may be able to suppress vocalization, but they cannot otherwise control it. Although their brains have homologues of the principal brain regions implicated in human language functions, Broca's and Wernicke's areas, these are responsible in other species for completely different activities (hand-eye coordination, for instance). A reorganization of the brain had to take place to get to the human configuration that underlies our language abilities, and that is not merely a matter of the brain's getting bigger.

Think of language as a tool that has its own structure and that we use to connect sound (or gesture) with meaning. In those terms, we find at least limited parallels to the linguistic form of sound and of meaning in other animals. What we do not find is any parallel at all to the tool itself. No other animal seems to employ anything like a discrete combinatorial system, so it is very hard to know where to look for precursors of the most important parts of language—phonology, morphology, and especially syntax.

On the analogy of the primates we can observe today, premodern hominids must have had some of the basic prerequisites for language. While their vocal tracts may not have been permanently adapted for speech as we find it now, these animals were more flexible than they are usually given credit for, and were probably able to vocalize with a fairly substantial (if limited) range of sounds.

Furthermore, even in the absence of speech they probably still had a fairly elaborate and articulated capacity for thought, involving the identification of specific individuals and the conceptualization even of asymmetric relations among these individuals. Work on the cognitive organization of nonhumans in the years since the decline of a radical behaviorist view of animals convinces us of that. We must conclude that our ancestor hominids on the threshold of language were capable of a certain degree of rational thought, and a certain degree of speech-like vocalization, even though there was no direct connection between the two.

The most plausible scenario (at least in my estimation) for how we arrived at modern language is the one provided by Derek Bickerton. He believes that the development of language involved two distinct stages: first, our premodern hominid ancestors, on the basis of the capacities discussed above, originated a limited form of communication that Bickerton calls protolanguage. It was communication on the basis of an open-ended vocabulary of individual spoken (or perhaps signed) words, but with essentially none of the combinatory organization that provides modern language with its greatest strengths. This protolanguage later grew into what we know today as language, along the following lines:

1. At a period long prior to the pongid-hominid split, there developed among primates a social calculus that included abstract representations of actions and their participants (thematic roles); these were preserved in episodic memory.
2. Over two million years ago, a structureless protolanguage developed in the hominid line.
3. For the next two million years, protolanguage continued but could not develop into true language because the level of signal coherence achievable by the human brain was too low to map the thematic structures of the social calculus onto utterances.
4. Within the last two hundred thousand years, an adequate coherence level was achieved in a group that became the ancestors of modern humans, causing the emergence of basic phrasal-clausal structure.

5. Crossing the coherence threshold unleashed a cascade of consequences, including the development of grammatical morphology and parsing algorithms that were incorporated into the human genome by Baldwinian evolution.

As should be obvious, this scenario is largely speculation — but fairly plausible speculation. Essential to Bickerton's view is the idea that the origin of language has two principal stages: the development of protolanguage, and its conversion into something more like modern language. Protolanguage would be a system whose speakers have a certain vocabulary but no real principles of syntactic combination. This sounds rather like animal call systems, but it is somewhat more interesting.

According to Jackendoff, a significant step in the transition from fixed calls to protolanguage was the separation of meaningful cries from specific situations. A vervet's eagle call means something like "Look out for the eagle!" — it cannot mean "Anybody seen an eagle lately?" or "That bird looks like an eagle (but isn't)" or anything else. Think of an infant's first words, though: "kitty" can mean "Look at the kitty," "Where's the kitty?" "Come here, kitty," "That looks like a kitty," and so on. The words are independent of the situation, and this constitutes an important difference between (proto)language and animal calls.

Further, even at the earliest stage, babies have some words that refer to particular *individuals* ("mommy," names) and some that refer to *types* ("kitty," "car"). This distinguishing between proper and common noun references is another feature lacking in animal calls. Animals certainly recognize specific individuals as opposed to types, but the distinction is not reflected in their communication. It is, however, present in (and essential to) real language — even protolanguage.

We can see that roughly this stage of sophistication is reached in the experiments that try to teach language to apes. The systems these animals develop have both common and proper nouns, and they can use their signs in a way that is at least partially independent of specific situations. They do not have equivalent symbols in nature, but the cognitive capacity to support them seems to be potentially present in a number of nonhuman species.

Actually, fully modern language has elements that are reminiscent of the kind of nonsyntactic character Bickerton and Jackendoff attribute to protolanguage. Consider words Jackendoff notes as "associated with sudden high affect, for instance *ouch! dammit! wow!* and *oboy!*" And then there are *sssh! psst!* and *hey!* Of particular interest are *hello* and *goodbye, yes* and *no,* which have fairly structured semantics and conditions of occurrence but no associated syntax.

So the idea is that a class of meaningful symbolic vocalizations emerged —quite possibly in association, at least initially, with nonvocal gesture. This last point does not really matter, although perhaps it bears on how language eventually came to be connected with Broca's area, a region that in other primates has something to do with hand-eye coordination, especially in connection with putting food into the mouth.

Once vocalizations of this sort are present, the variety of things that it might be useful to communicate—and that can be communicated in this way—is fairly large, so the vocabulary grows. But if each "word" is just a logically simple, unanalyzable whole, the number of these items that people can keep track of and use distinctly is limited. So in association with the expansion of the vocabulary, it is useful to treat individual words as made up of recurring parts. Combining only a few such parts in different ways yields a great many possible words. Ten basic items (syllables, for instance), combined in sequences of three at a time, yield a thousand distinct "words."

Here we have the beginning of phonology, an aspect of language that no ape ever seems to catch on to. Perhaps that cognitive gap underlies the fact that even with a great deal of training and a great many opportunities to observe, ape vocabularies seem to top out at a few hundred signs. Probably phonological structure was initially a matter of a limited range of syllables rather than phonemes. Syllables seem to be a very natural unit, with analysis into segment-sized units coming later.

We are assuming that our early hominids had fairly structured thoughts, and in particular that a single thought could involve, for instance, both grooming and individuals doing the grooming and getting groomed. Such a complex thought might well call forth more than one word, so we get words of protolanguage combined with one another. The combination, however, is largely illusory. At this level of analysis, it is simply an unstructured conjunction of a number of words that are prompted more or less simultaneously by a single, complex idea—rather like semantic soup.

Just what a given combination might mean is something that would have to be reconstructed from pragmatic knowledge. In that respect, it is similar to some constructions in modern languages. Consider the formation of compound nouns in English: a *snowman* is a man made of snow (or at a ski resort, it might be the man who operates the snow-making machine), the *mailman* and the *milkman* bring the mail and the milk, but the *garbage man* takes the garbage away, a *first baseman* is the baseball player who plays at first base, and we also have the *barman*, the *craftsman, oarsman, swords-man, cave man, frogman,* and *family man* (among many others even when we confine our attention to combinations of *man* with another noun). In each of these cases, the relation between *man* and the other element of the compound is not specified as a matter of grammar, only by whatever connection seems most plausible on the basis of our knowledge of the world.

This kind of combination might look like syntax, but when we remember how syntax really works, we will not take that possibility too seriously. We can ignore everything about displacement, D-structure, constraints on displacement, and the like: this kind of combination does not even involve hierarchical structure and phrases. It still allows for the expression of a somewhat wider class of meanings than just single signs, though, so its emergence can be seen to have enhanced the communicative usefulness of the protolinguistic system.

We can add a few principles of ordering at this point, giving more significance to the sequence in which the components of a complex meaning are produced. For example, it seems that a rule such as "agent first" is a general ordering regularity in languages (including those with little or no other structure, such as pidgins), even though it is not universal and is violated by specific constructions in most languages.

This, then, is protolanguage: a communication system based on a growing (indeed, open-ended) vocabulary of individual words, each built up from a smaller inventory of sound units and produced as the expression of a structured thought. We can equate this kind of system, at least in gross terms, with the level achieved by language-trained apes, although it appears that the apes lack any appreciation of the phonological dimension of their system. Some people—especially those who want to connect ape language experiments with human language—would say that children at the two-word stage basically control a system of this sort. Others argue that children at that point already have rather full command of syntax, but because of other limitations on their production cannot put together sentences long enough to make that control obvious.

We can also associate this kind of structure with that of the pidgins which form among groups of people thrust together without any common language. Bickerton argues that agrammatic aphasics, and perhaps individuals deprived of language input during the sensitive period for language acquisition (such as Genie in Chapter 6), have a level of development of roughly this sort.

What is centrally lacking from protolanguage is the kind of combinatorial system that lies at the heart of the syntax, morphology, and phonology of modern spoken languages. I do not propose to speculate here on the developments that led from protolanguage to full language through the rise of structure of this sort. A number of logically possible paths are outlined in the literature, and new ones seem to appear all the time.

What matters in the context of this book is an appreciation of how far other species still must go to arrive at something like human natural language, despite the obvious potentialities of the cognitive structure we have in common with other primates. No other primate functions communicatively in nature even at the level of protolanguage, and the vast gulf of discrete, recursive combinability must still be crossed to get from there to the language capacity inherent in every normal human. We seem to be alone on our side of that gulf, whatever the evolutionary path we may have taken to get there.

What's So Great about Language?

As we come to the end of this book, I propose to lay out an argument that brings a number of strands together, a conclusion about just what constitutes the core strength of human natural language. It goes against the grain of traditional discussions of this matter. Nevertheless, I believe that it is warranted on its own, and that it derives further strength from what we learn by studying the cognitive capacities of other animals.

I contend that the most important aspect of language by far is not the capacity to represent meanings by arbitrary symbols, but the syntactic system. That is what allows us to combine those symbols into arbitrarily complex constructions whose novel meanings are (largely) a compositional function of the meanings of the parts and their manner of combination. It is that possibility that gives language users their strong advantage.

Surely language is an enormously meaningful development in the history of our species. Because we can talk about what we have done, or might do, or what others have done, we can learn from experience, plan, and accumulate the intellectual and technological capital we have. Without language, whatever other intellectual advantages we might have would not be communicable to others, and the understanding and achievement of each individual human being would die with that person. We do not run the world because of our ability to dominate other species physically. It is the possession of language that has given humankind a huge advantage in evolutionary competition with other species for dominion over the earth.

What is it, then, that makes language so important? Some have argued that what is essential to the power of language is its nature as a *semiotic* system: that is, the fact that words are abstract signs. This goes back to the views of Ferdinand de Saussure and Charles S. Peirce in the nineteenth century, for whom a language was essentially a system of signs. The kind of "system" they implied, however, was only marginally more than a mere *collection* of signs (arbitrary sound/meaning pairs). What was systematic for Saussure or Peirce was the relation one sign bears to others, not something above and beyond the signs themselves.

In line with this view, most traditional "philosophy of language" is concerned with what exactly words can and do mean, and how it is that they manage to mean anything at all. The focus is on the nature of symbolic relations between the outward forms of words (sounds, or marks on a page) and their meanings. Even Edward Sapir, normally high in my personal pantheon of thinkers about language, says that "the essence of language consists in the assigning of conventional, voluntarily articulated, sounds, or of their equivalents, to the diverse elements of experience." I do not think this can be right. That is, I do not think that it is by virtue of its ability to deploy arbitrary symbols that our language derives its most significant force.

In earlier parts of this book we have seen that some animals evidently have the cognitive capacity to control a system of abstract, symbolic signs. This assertion is rather strong, in the overall context of historical discussion, so we should examine it more closely.

What do we mean when we say the animals in question have a system of symbolic signs? Actually, the evidence is robust. What we look for is evidence that animals can use arbitrary signs in the way speakers of human languages do—to refer to concepts, objects, and relations in the world. We need to show symbol use that meets at least the conditions mentioned in Chapter 10. It should be *noninstrumental* in the sense that the symbols are genuinely used to *refer* to something, not simply as a means for carrying out some specific action or getting something. It should be *displaced*, in the sense that the symbols can be used to refer to things that are not necessarily present in the environment when the symbol is used. And it should be *noniconic*, in the sense that the symbols do not directly represent the things they stand for.

As with the attempt to define human language by a set of specific properties, any effort to set up a list of criteria by which symbolic expression can be identified externally runs the risk of contradiction or trivialization. Still, these conditions seem like a reasonable starting point, and they correspond to the spirit of most discussions in the literature of what it would take for an animal's behavior to involve truly symbolic reference.

Most of the ape utterances of which we have any serious record are certainly instrumental: they are ways to get food or treats. Sue Savage-Rumbaugh stresses the fact that Kanzi asks to have other people chase each other but, as Joel Wallman suggests, either bonobos have a more diffuse sense of reward or Kanzi is simply "a chimpanzee with strange tastes." In most cases, Kanzi's utterances too are intended to get something. Kanzi's ability to use manual gestures and his keyboard, and also to recognize spoken English words, is extraordinarily impressive; but the utterances he produces still do not involve comments on things that have (or have not, or might have) happened, questions, and so on. Reported instances of signing and keyboard pressing (or in Alex's case, vocalizing) when alone, which abound in the literature but remain anecdotal, are minimal exceptions.

On the other hand, Kanzi's behavior in testing situations makes it reasonably obvious that he does, in fact, understand the notion that abstract symbols such as those on his keyboard and sounds that he hears "stand for" something. This point was already evident from Sherman and Austin's ability to classify symbols in terms of properties of their referents. One can argue that, overall, Kanzi wants to do well on tests so as to get more generalized rewards (praise, love, whatever), but that does not suffice to provide a locally instrumental interpretation for what he does when, say, he correctly identifies an object from its symbol (or vice versa).

Alex's behavior, similarly, does not seem subject to a purely instrumental interpretation. Alex makes statements on the basis of a rather diffuse desire to be part of the conversation, to interact. And while many of the things Alex identifies are things he wants, and when he gets the right answer he gets a reward, the utterance itself is much less directly related to the obtaining of gratification than "Please Irene give [Alex] nut."

The evidence here favors an interpretation on which at least Kanzi and Alex can use their symbols in noninstrumental ways, although it has generally proved rather hard for humans to get other experimental subjects to do as much.

With respect to displacement, the evidence is rather limited. Both

Kanzi and Alex, and to some extent a few of the chimpanzees, express their desire for things that are not in fact presently visible. This is still a long way from an animal's telling us a story about another animal she met last week, or even about the one we both know is in the next room. The ability of any animal to use symbols in a way that is completely independent of the presence of what they refer to has not been solidly established to date.

As for the noniconicity of the symbols animals can be taught to use, that is not seriously in doubt. The lexigrams employed by Kanzi and other primates in similar experiments are wholly noniconic, so if we accept that the animal has a sense that the lexigram is a sign for something, it is clearly a noniconic sign.

In the sign language experiments, many of the signs the animals use do represent their referent iconically. Consider "open" for instance, manifestly an opening gesture; or pointing gestures; or touching parts of the body that are to be attended to, and so forth. The fact that some chimpanzees indicate tickling of someone else by making the tickling sign on that individual is a matter of making it more iconic in a way that does not correspond to what happens with signs in any natural signed (human) language, so far as we know. This fact does not compromise the observation that other signs used by the chimpanzees are undoubtedly learned, arbitrary associations. The "learned" part of this formulation is presumably important, since the learning of novel symbols is what supports the open-ended quality of our vocabulary. Similarly, the apes in the language training experiments can obviously learn new symbols (gestures, keystrokes, or—at least receptively—spoken words). And of course so can Alex.

In Alex's case, the symbols are completely noniconic; they are vocal in nature but indicate aspects of the world having no nonarbitrary connection with sound. The same is true, of course, of the spoken words Kanzi understands, even though he cannot reproduce them directly himself.

A final property we might look for as evidence for the symbolic nature of animals' signs, vocalizations, key presses, and the like is the existence of some larger system that goes beyond a simple inventory. In the case of the apes' keyboards, the way the signs are presented makes it unlikely that such structure will be there; the experimenters have taken pains to make the symbols as independent of one another as possible. We have no evidence that the capacity to learn phonology—a combinatory system that underlies a set of symbols, but whose elements are meaningless in themselves—has ever been seriously tested for.

In the case of the signing chimpanzees, structure of this sort surely exists in their input (recall the role of handshape in pronominal signs), but despite the systematicity of this aspect of human signers' productions, we have no reason to believe that any ape has ever picked up on this or any other systematic aspect of the internal structure of signs.

With Alex, it is obvious that he has developed a structure among the concepts represented by the words he knows. Colors and shapes are superordinate categories of this sort. To a large extent, his system is the result of quite specific training, but regardless of where it comes from, his set of symbols clearly has some internal organization.

The question of whether he treats the individual words themselves as the products of some sort of combinatorial system is more complex. Pepperberg reports that he spontaneously produces variations on spoken words he has learned. She has taken advantage of this in his training: when he produced "grain" soon after learning the word "grey," he was given some parakeet seed that was identified to him as "grain," and this word entered his vocabulary. Pepperberg interprets these innovations as based on a phonemic system, but considerably more study is required before we can actually justify that conclusion. It would be fascinating, for instance, if we determined that Alex finds "three" and "key" to have something in common (their vowel), but at the moment we have no reason to believe any such thing.

Some of the questions posed above have not in fact been addressed in research to date, and it is sensible to defer negative conclusions until the relevant experiments are done. But it does not seem unfair to note that, despite the systematic phonological and morphological properties of signed languages like ASL that have been present in the input in several experiments, we have no basis for believing that this structure is learnable (for apes) in anything like the spontaneous way human children acquire it.

To summarize, we do have a certain amount of evidence for the ability of animals to learn arbitrary symbolic expression, even if some of them actually use it only to get foods they want to eat. That in itself is remarkable, for the kind of symbolic reference we can induce in the laboratory is an ability no animal apparently displays *without* explicit training; so far as we know, no animal in nature uses a system that involves abstract symbolic reference.

Or if they do, we have no evidence for it. It is logically possible that symbolic systems exist in nature, which we have just failed to noticed; or that the animals systematically hide their symbol use from humans (as Calvin's tiger Hobbes always turns into a stuffed toy the minute some adult is present). These are not likely alternatives, though — and if some animals do in fact utilize symbols (without our knowing), that clearly has not given them the capacity to challenge human influence within their environment.

Enter Chomsky's argument. We have to assume that if a group of animals has some capacity which, at least potentially, could confer a significant evolutionary advantage, it is extremely unlikely that this capacity will go unexploited over evolutionary time. As a result, when we find that animals of a given species fail to display some specific ability in nature, that provides us with reason to conclude either that the ability in question is beyond the capacity of these animals, or else that the ability does not in itself confer an evolutionary advantage.

From these considerations, it seems we must conclude that whatever makes language such a tremendously powerful tool must be something other than symbolic reference alone. The relevant property must be something to which humans have access but nonhumans do not; and which (perhaps *combined with* the capacity for symbolic reference) yields the richly expressive system humans have used to create culture and take over the earth (if not yet the entire universe . . .).

An obvious candidate for such a property is syntax. It is syntax that allows us to combine a relatively small fixed vocabulary of symbolic expressions into a literally unbounded range of different, novel utterances that anyone who speaks the same language can produce and understand. This ability is what takes us from protolanguage to language.

And it is precisely syntax, in this open-ended, unbounded, recursive, fully productive sense, that no ape, parrot, dolphin, or any other animal has been shown to control — either naturally, or with arbitrarily extensive training. Savage-Rumbaugh, among others, has complained that linguists seem to be obsessed with syntax, but the reason is not just perversity. Her own extremely important work with Kanzi has shown us that symbolic expression is within the cognitive capacity of other species. But no such result exists for syntactic organization. It seems entirely reasonable to say that the essence of language is to be found precisely here, in the flexibility of expression that results from syntactic combination.

As we have seen, many key aspects of the syntactic systems of natural languages are significantly underdetermined by the evidence available to the language learner. These properties are nevertheless found universally —

not just in some languages, but apparently in all, and not just after exposure to education, but in unwritten languages as well. The most plausible assumption, to my mind, is that these are aspects of the cognitive organization human children bring to the task of language acquisition. Since no other species seems to be able to do the same, it appears that the syntactic principles of Universal Grammar are a part of specifically human biology.

Syntactic combination thus seems to be the key to the efficacy of language, a uniquely human faculty provided to us by our biology. Where it came from and how it evolved are fascinating questions, but not questions that need to be answered in order to attest to the existence of syntax as a part of our nature. Humans are indeed unique—no more so than every other species, but not less so either. "The fox has many tricks, the hedgehog only one: one good one." And our special trick is a really good one.

Notes

1 Animals, Language, and Linguistics

Page **3**—Review of books about apes: Douglas H. Chadwick, Our unfortunate cousins, *New York Times Book Review,* 11 December 1994.

Page **4**—Response to Chadwick's review: David Pesetsky, How to tell the apes, *New York Times Book Review,* 25 December 1994, p. 23.

Page **7**—Actual number of human genes: *Science* 300:1484 (2003)

Page **7**—Proportion of genetic material devoted to the brain: *Science* 291:1188 (2001).

Page **14**—Russell on dogs and language: Russell 1948, p. 74.

2 Language and Communication

Page **16**—Communication among bacteria: Andrew Pollack, Drug makers listen in while bacteria talk, *New York Times,* 27 February 2001.

Page **18**—Evolution of communication: For an extensive and detailed discussion, see Hauser 1996.

Page **21**—Design Features for Language: Hockett 1960. Hockett's paper circulated in a number of different versions, with the precise list of design features changing slightly over time. The precise set discussed here is not identical to any single version of the paper; it is a starting point for discussion of interesting issues, not an exposition of Hockett's specific views.

Page **21**—Evolution of physical specialization for speech: Lieberman 1984.

Page **21**—Less-common sensory modalities: Hughes 1999.

Page **27**—Language-particular forms of animal sounds: See www.georgetown.edu/cball/animals/.

Page **31**—Language as made for lying: Sturtevant 1947, p. 48.

Page **32**—The piping plover's trick: Ristau 1991.

Page **32**—Machiavellian intelligence: Byrne and Whiten 1992.

Page **35**—Signature characteristics: Bradshaw 1993.

3 On Studying Cognition

Pages **41ff.**—The story of Clever Hans: Pfungst 1911.

Page **43**—Clever Hans's real abilities: Cited in Pfungst 1911, p. 5.

Page 44 — Morgan's canon: Morgan 1894, p. 53.

Page 47 — Theory of mind in apes: Povinelli 2000.

Page 48 — Deceptive watchbirds: Munn 1986.

Page 52 — See Anderson and Lightfoot 2002 for a more comprehensive discussion of the components of human language and their relation to human biology and cognition.

Page 53 — Nicaraguan Sign Language: Kegl, Senghas, and Coppola 1999.

Page 54 — Spontaneous development of signing in deaf children: Goldin-Meadow and Mylander 1990.

Page 54 — Creole development: For extensive discussion, see the papers in DeGraff 1999.

Page 55 — Dissociation of language and general cognitive development: Curtiss 1988.

Page 55 — Specific Language Impairment: Leonard 1998.

Page 55 — Williams syndrome: Bellugi et al. 1993.

Page 55 — Christopher the linguistic savant: Smith and Tsimpli 1995; learning BSL: Morgan et al. 2002.

Page 59 — Japanese advertising copy: Examples are collected at www.engrish.com.

4 The Dance "Language" of Honeybees

Page 64 — Bee dances as a language: Gould and Gould 1995, p. 59.

Pages 65ff. — The facts discussed in this section derive from several sources. Although the classic description is that of von Frisch 1967, several recent works are highly readable and update his account in important ways. Notable are Gould and Gould 1995, Seeley 1995, and Wenner and Wells 1990. The last takes a highly skeptical view of von Frisch's results (and methods), a matter discussed from the other side in chapter 4 of Gould and Gould 1995. My discussion cannot possibly do justice to all that is known about bees and their dances. Dyer 2002 provides an updated and broader survey, with references.

Page 65 — Aristotle citation: Von Frisch 1967, p. 6.

Page 66 — Bee dance as a language: Lubbock 1874, p. 160.

Page 66 — Odor as the key to foragers' success: Maeterlink 1901; Lineburg 1924.

Page 67 — Mystery of communication in the dance: Francon 1939, p. 143.

Page 68 — Von Frisch and the Nazis: Gould and Gould 1995, p. 58. Actually, von Frisch's relations with the Nazi authorities and his expressed opinions during the period 1941–45 were nowhere near as unambigiously heroic as his postwar rehabilitation would suggest. These matters, explored in some depth in Deichmann 1996, pp. 171–200, do not bear directly on the scientific issues at stake here.

Page 70 — The economics of energy and foraging: Seeley 1995.

Page 70 — Information transfer in the dance: Michelson 1999.

Page 73 — Role of sound in the dance: Michelson 1999.

Page 75 — Trembling dance: von Frisch 1967, p. 282, quoting from one of his earlier papers.

Page 76 — Cricket chirping not "language": Gould and Gould 1995.

Page 80 — Discrete combinatioral systems: Pinker 1994, p. 84.

Page 83 — Critics of the dance language hypothesis: Wenner and Wells 1990. Wenner 2002 attempts to revive earlier criticisms of the claim that dance (as opposed to odor) communicates the location of a food source to other potential foragers. However, he does not really rebut the arguments that have been offered in favor of this view since the appearance of the earlier book.

Page 88 — A thorough discussion of some of the issues in insect navigation, not limited to bees and their dances, is Gallistel 1998.

5 Sound in Frog and Man

Page 92 — Bug detectors in the frog's visual system: Lettvin et al. 1959.

Page 94 — South African clawed frogs: Kelley and Tobias 1999.

Page 98 — Gerhardt and Huber 2002 provide further information about the precise mechanisms by which frogs of various kinds produce sound.

Page 98 — Wilczynski, Zakon, and Brenowitz 1984 detail the vocal communication of the spring peeper, with attention to the underlying neurophysiology.

Page 100 — A summary of this early work on the bullfrog is in Capranica 1965.

Page 107 — Those interested in delving deeper into acoustic phonetics are referred to Ladefoged 1996.

Page 110 — Speech is special: Liberman 1982. Papers by Liberman and his colleagues relevant to the material of this section are collected in Liberman 1996.

Pages 115ff. — Motor theory of speech perception: Again, see the papers in Liberman 1996.

Page 124 — Cognitive importance of distinctive features: Jackendoff 1994.

Page 125 — Two olfactory systems in mice: *Science* 299:1196–1201 (2003).

6 Birds and Babies Learning to Speak

Page 133 — Chickadee call structure: Hailman, Ficken, and Ficken 1985.

Page 135 — Song versus calls: Marler 1999. Points cited later in the chapter also derive from this classic discussion.

Page 142 — Neurophysiology of birdsong: Nottebohm 1999.

Page 143 — Motor theory and birdsong perception: Williams and Nottebohm 1985.

Page 145 — Notions of song "learning" across a variety of species are surveyed in Boughman and Moss 2003.

Page 147 — Brain nuclei common to oscines, hummingbirds, and parrots: Jarvis et al. 2000, p. 632.

Page 155 — Relation between seasonal learning and neurogenesis: Hauser 1996, pp. 144ff.

Page 157 — Human language organ: Pinker 1994; Anderson and Lightfoot 2002.

Page 161 — Acquisition of human language: Our knowledge of the earliest stages of the child's path to language has expanded greatly since the early 1990s. The account here is drawn from Anderson and Lightfoot 2002 and relies heavily on Jusczyk 1997, de Boysson-Bardies 1999, and Kuhl 1999. De Boysson-Bardies's book, in particular, provides a readable introduction to much of this research.

Page 161 — Manual babbling by deaf babies: Petitto and Marentette 1991.

Page 164—Genie: Curtiss 1977. Curtiss 1988 provides a somewhat broader survey of classic cases relevant to language learning when exposure to language is lacking during the sensitive period for acquisition. Emmorey 2002 surveys the literature dealing with the special case of hearing-impaired children who gain access to (signed) language relatively late in life.

7 What Primates Have to Say for Themselves

Page 169—First observations of alarm calling in vervets: Struhsaker 1967.

Page 171—Squirrel alarm calls based on urgency: Macedonia and Evans 1993.

Page 171—See Cheney and Seyfarth 1990 for a wealth of information about communication and much more in vervet monkeys.

Page 172—Unreliable signals in ground squirrels: Hare and Atkins 2001.

Page 172—Prairie dog alarm calls: Slobodchikoff 2002.

Page 175—Combining of calls in forest monkeys: Zuberbühler 2002.

Page 180—Limbic system in primate vocal production: Deacon 1992.

Page 182—The vocal repertoire of the ring-tailed lemur, based on studies of lemurs in a study colony at Duke University, is described in Macedonia 1993. Included are details of the acoustics of these calls and the circumstances under which they have been observed.

Page 184—Chemical signals in lemurs: Kappeler 1998.

Page 185—Chemically assisted theft in tropical ants: Breed et al. 1990.

Page 191—Functional interpretation of primate calls: Owren and Rendall 2001.

Page 195—Chimpanzees' inability to produce deceptive vocalizations: Goodall 1986, p. 125.

8 Syntax

Page 214—Requirements of verbs for specific arguments: Pinker 1994, pp. 112ff.

9 Language Is Not Just Speech

Page 234—The linguistic analysis of signed languages, especially ASL, has generated a huge literature. An early introduction to this work, which retains much of its value, is Klima and Bellugi 1979. A more recent collection updating many of their points is Emmorey and Lane 2000. The impact of modality on the structure of language in general is addressed in Emmorey 2002 and Meier, Quinto, and Cormier 2002. I have also benefited greatly from Perlmutter 1991 and a 1996 presentation at the American Association for the Advancement of Science by Richard Meier, many of whose examples I use here.

Page 234—First treatment of ASL in linguistic terms: Stokoe 1960.

Page 236—Martha's Vineyard signing community: Groce 1985.

Page 257—I am grateful to Susan Fischer for suggesting these examples and for confirming the judgments here with native signers.

Page 260—Home sign systems: Goldin-Meadow and Mylander 1990, Goldin-Meadow 2003.

Page 260—Development of creoles: DeGraff 1999, among others.

Page 260—Emergence of Nicaraguan Sign Language: Kegl, Senghas, and Coppola 1999.

Page 260—Effects of modality in signed languages: Meier, Quinto, and Cormier 2002.

10 Language Instruction in the Laboratory

Page 269—The initial substantive report on the Gardners' work with Washoe is Gardner and Gardner 1969. Subsequent papers expand the picture of this project and its results, and a collection of relevant papers is found in Gardner, Gardner, and van Cantfort 1989. Wallman 1992 analyzes the ape language projects dealt with in this chapter through Savage-Rumbaugh's early work with Kanzi, and he finds serious defects in all of them. His negative assessment may be overstated at times, but from the point of view of experiment psychology his objections are generally cogent.

Page 269—Washoe's later years: Fouts 1997.

Page 276—Terrace 1979 gives a full report on the Nim project.

Page 277—Grammatical sophistication in early childhood: Hirsch-Pasek and Golinkoff 1996 discuss some innovative experiments on the grammatical abilities of very young children.

Page 281—Nim's signing as imitation: Terrace et al. 1979.

Page 282—Nim's more conversational signing: O'Sullivan and Yeager 1989.

Pages 283ff.—Chantek project: Miles 1990.

Page 286—Plural of "anecdote": Bernstein 1988, p. 247.

Page 287—Sarah project: Premack and Premack 1972.

Page 287—Lana project: Rumbaugh 1977.

Page 288—Sherman and Austin: Savage-Rumbaugh 1986.

Pages 289ff.—Kanzi: Kanzi and his accomplishments have been described at length in two books addressed to general audiences: Savage-Rumbaugh et al. 1986 and Savage-Rumbaugh, Shanker, and Taylor 1998. In my opinion, the two books suffer from a combination of exaggeration and defensiveness, but both offer fascinating information and perspective.

Page 291—Kanzi's development compared to that of a human child: Savage-Rumbaugh et al. 1993.

Pages 300ff.—Alex: Pepperberg 2000.

Page 301—The mechanisms of sound production in mynah birds, and the acoustic relations between their sounds and the human speech that it sounds like to us, are described by Klatt and Stefanski 1974. Pepperberg 2000, chapters 15 and 16, provides details about the corresponding issues in the speech of parrots such as Alex.

11 Language, Biology, and Evolution

Page 307—Chomsky's argument about flightless birds: Chomsky 1980, p. 239.

Pages 309ff.—Most of the discussion in this section derives from work by linguists such as Carstairs-McCarthy 2001 and Jackendoff 2002, or from scholars who have collaborated actively with linguists. Hauser, Chomsky, and Fitch 2002 presents many of these matters from one particular point of view.

Page 311 — Vocal tract shape as a preadaptation: Carstairs-McCarthy 2001.

Page 311 — Lowering of the larynx in other species: Fitch 2002.

Page 312 — Vocal abilities of Neanderthals: Lieberman 1984.

Page 314 — Bickerton's account of the emergence of human language: Bickerton 1990, 1995, 2000.

Page 315 — Baldwinian evolution is the enhancement through natural selection of the ability to learn certain advantageous skills or behaviors. The advantage conferred by useful behavior produces selectional pressure in favor of whatever genetic basis supports it. This differs from standard Darwinian evolution not in terms of mechanisms, but because what evolves is the tendency to acquire a behavioral trait (flying, speaking) rather than a physical characteristic (wings, shape of the vocal tract).

Pages 315ff. — Relation of protolanguage to modern linguistic forms: Jackendoff 2002.

Page 319 — Sapir on the essence of language: Sapir 1921, p. 11.

⤙ References ⤚

Anderson, Stephen R., and David W. Lightfoot. 2002. *The Language Organ: Linguistics as Cognitive Physiology.* Cambridge: Cambridge University Press.

Bellugi, Ursula, S. Marks, A. Bihrle, and H. Sabo. 1993. Dissociation between language and cognitive functions in Williams syndrome. *Language Development in Exceptional Circumstances,* ed. D. Bishop and K. Mogford. Hillsdale, N.J.: Lawrence Erlbaum Associates.

Bernstein, I. S. 1988. Metaphor, cognitive belief, and science. *Behavioral and Brain Sciences* 11: 247–248.

Bickerton, Derek. 1990. *Language and Species.* Chicago: University of Chicago Press.

Bickerton, Derek. 1995. *Language and Human Behavior.* Seattle: University of Washington Press.

Bickerton, Derek. 2000. How protolanguage became language. *The Evolutionary Emergence of Language,* ed. C. Knight, M. Studdert-Kennedy, and J. R. Hurford, pp. 264–284. Cambridge: Cambridge University Press.

Boughman, Janette Wenrick, and Cynthia F. Moss. 2003. Social sounds: Vocal learning and development of mammal and bird calls. *Acoustic Communication,* ed. A. M. Simmons, A. N. Popper, and R. R. Fay, pp. 138–224. New York: Springer Verlag.

Bradbury, Jack W., and Sandra L. Vehrencamp. 1998. *Principles of Animal Communication.* Sunderland, Mass.: Sinauer Associates.

Bradshaw, Gary. 1993. Beyond animal language. *Language and Communication: Comparative Perspectives,* ed. H. L. Roitblat, L. M. Herman, and P. E. Nachtigall, pp. 25–44. Hillsdale, N.J.: Lawrence Erlbaum Associates.

References

Breed, Michael D., P. Abel, T. J. Bleuze, and S. E. Denton. 1990. Thievery, home ranges, and nestmate recognition in *Ectatomma ruidum*. *Oecologia* 84: 117–121.

Byrne, R. W., and A. Whiten. 1992. Cognitive evolution in primates: Evidence from tactical deception. *Man* 27: 609–627.

Capranica, Robert R. 1965. *The Evoked Vocal Response of the Bullfrog*. Cambridge, Mass.: MIT Press.

Carstairs-McCarthy, Andrew. 2001. Origins of language. *The Handbook of Linguistics*, ed. M. Aronoff and J. Rees-Miller, pp. 1–18. Oxford: Blackwell.

Cheney, Dorothy L., and Robert M. Seyfarth. 1990. *How Monkeys See the World*. Chicago: University of Chicago Press.

Chomsky, Noam. 1957. *Syntactic Structures*. The Hague: Mouton.

Chomsky, Noam. 1980. *Rules and Representations*. New York: Columbia University Press.

Curtiss, Susan. 1977. *Genie: A Psycholinguistic Study of a Modern-Day "Wild Child."* New York: Academic Press.

Curtiss, Susan. 1988. Abnormal language acquisition and the modularity of language. *Linguistics: The Cambridge Survey*, ed. F. J. Newmeyer. Vol. 2, pp. 96–116. Cambridge: Cambridge University Press.

de Boysson-Bardies, Bénédicte. 1999. *How Language Comes to Children*. Cambridge, Mass.: MIT Press.

Deacon, Terrence W. 1992. The neural circuitry underlying primate calls and human language. *Language Origin: A Multidisciplinary Approach*, ed. J. Wind et al., pp. 121–162. Amsterdam: Kluwer Academic.

DeGraff, Michel, ed. 1999. *Language Creation and Change: Creolization, Diachrony and Development*. Cambridge, Mass.: MIT Press.

Deichmann, Ute. 1996. *Biologists under Hitler*, trans. Thomas Dunlap. Cambridge, Mass.: Harvard University Press.

Dyer, Fred C. 2002. The biology of the dance language. *Annual Review of Entomology* 47: 917–949.

Emmorey, Karen. 2002. *Language, Cognition and the Brain: Insights from Sign Language Research*. Mahwah, N.J.: Lawrence Erlbaum Associates.

Emmorey, Karen, and Harlan S. Lane, eds. 2000. *The Signs of Language Revisited: An Anthology to Honor Ursula Bellugi and Edward Klima*. Mahwah, N.J.: Lawrence Erlbaum Associates.

Fitch, W. Tecumseh. 2002. Comparative vocal production and the evolution of speech: Reinterpreting the descent of the larynx. *The Transition to Language,* ed. A. Wray, pp. 21–45. Oxford: Oxford University Press.

Fouts, Roger. 1997. *Next of Kin: What Chimpanzees Have Taught Me about Who We Are.* New York: William Morrow.

Francon, Julien. 1939. *The Mind of the Bees.* London: Methuen.

Gallistel, C. R. 1998. Symbolic processes in the brain: The case of insect navigation. Methods, models and conceptual issues. Vol. 4 of *Invitation to Cognitive Science,* ed. D. Scarborough and S. Sternberg, pp. 1–51. Cambridge, Mass.: MIT Press.

Gardner, R. Allen, and Beatrix T. Gardner. 1969. Teaching sign language to a chimpanzee. *Science* 165: 664–672.

Gardner, R. Allen, Beatrix T. Gardner, and Thomas E. van Cantfort, eds. 1989. *Teaching Sign Language to Chimpanzees.* Albany: State University of New York Press.

Gerhardt, H. Carl, and Franz Huber. 2002. *Acoustic Communication in Insects and Anurans.* Chicago: University of Chicago Press.

Goldin-Meadow, Susan. 2003. *The Resilience of Language.* Philadelphia: Psychology Press.

Goldin-Meadow, Susan, and C. Mylander. 1990. Beyond the input given: The child's role in the acquisition of language. *Language* 66: 323–355.

Goodall, Jane. 1986. *The Chimpanzees of Gombe: Patterns of Behavior.* Cambridge, Mass.: Harvard University Press.

Gould, James L., and Carol Grant Gould. 1995. *The Honey Bee.* New York: Scientific American Library.

Griffin, Donald R. 2001. *Animal Minds.* 2nd ed. Chicago: University of Chicago Press.

Groce, Nora Ellen. 1985. *Everyone Here Spoke Sign Language: Hereditary Deafness on Martha's Vineyard.* Cambridge, Mass.: Harvard University Press.

Hailman, J. P., M. S. Ficken, and R. W. Ficken. 1985. The "chick-a-dee" call of *Parus atricapillus:* A recombinant system of animal communication compared with written English. *Semiotica* 56: 191–224.

Hare, James F., and Brent A. Atkins. 2001. The squirrel that cried wolf: Reliability detection by juvenile Richardson's ground squirrels (*Spermophilus richardsonii*). *Behavioral Ecology and Sociobiology* 51: 108–112.

Hauser, Marc D. 1996. *The Evolution of Communication.* Cambridge, Mass.: MIT Press.

References

Hauser, Marc D. 2000. *Wild Minds: What Animals Really Think.* New York: Henry Holt.

Hauser, Marc D., and Mark Konishi, eds. 1999. *The Design of Animal Communication.* Cambridge, Mass.: MIT Press.

Hauser, Marc D., Noam Chomsky, and W. Tecumseh Fitch. 2002. The faculty of language: What is it, who has it, and how did it evolve? *Science* 298: 1569–79.

Hirsch-Pasek, Kathy, and Roberta Michnick Golinkoff. 1996. *The Origins of Grammar: Evidence from Early Language Comprehension.* Cambridge, Mass.: MIT Press.

Hockett, Charles F. 1960. Logical considerations in the study of animal communication. *Animal Sounds and Animal Communication,* ed. W. E. Lanyon and W. N. Tavolga, pp. 392–430. Washington, D.C.: American Institute of Biological Sciences.

Hughes, Howard C. 1999. *Sensory Exotica.* Cambridge, Mass.: MIT Press.

Jackendoff, Ray S. 1994. *Patterns in the Mind.* New York: Basic Books.

Jackendoff, Ray S. 2002. *Foundations of Language.* Oxford: Oxford University Press.

Jarvis, Erich D., Sidarta Ribeiro, Maria Luisa Da Silva, Dora Ventura, Jacques Vielliard, and Claudio V. Mello. 2000. Behaviourally driven gene expression reveals song nuclei in hummingbird brain. *Nature* 406: 628–632.

Jusczyk, Peter. 1997. *The Discovery of Spoken Language.* Cambridge, Mass.: MIT Press.

Kappeler, Peter M. 1998. To whom it may concern: The transmission and function of chemical signals in *Lemur catta. Behavioral Ecology and Sociobiology* 42: 411–421.

Kegl, Judith A., Annie Senghas, and M. Coppola. 1999. Creation through contact: Sign language emergence and sign language change in Nicaragua. In DeGraff 1999.

Kelley, Darcy B., and Martha L. Tobias. 1999. Vocal communication in *Xenopus laevis.* In Hauser and Konishi 1999.

Klatt, Dennis H., and Raymond A. Stefanski. 1974. How does a mynah bird imitate human speech? *Journal of the Acoustical Society of America* 55: 822–832.

Klima, Edward, and Ursula Bellugi. 1979. *The Signs of Language.* Cambridge, Mass.: Harvard University Press.

Kroodsma, Donald, and Edward H. Miller, eds. 1996. *Ecology and Evolu-*

tion of Acoustic Communication in Birds. Ithaca, N.Y.: Cornell University Press.

Kuhl, Patricia K. 1999. Speech, language and the brain: Innate preparation for learning. In Hauser and Konishi 1999.

Ladefoged, Peter. 1996. *Elements of Acoustic Phonetics*. 2nd ed. Chicago: University of Chicago Press.

Leonard, Laurence B. 1998. *Children with Specific Language Impairment*. Cambridge, Mass.: MIT Press.

Lettvin, J. Y., H. R. Maturana, W. S. McCulloch, and W. H. Pitts. 1959. What the frog's eye tells the frog's brain. *Proceedings of the IRE* 47: 1940–59.

Liberman, Alvin M. 1982. On finding that speech is special. *American Psychologist* 37: 148–167.

Liberman, Alvin M. 1996. *Speech: A Special Code*. Cambridge, Mass.: MIT Press.

Lieberman, Philip. 1984. *The Biology and Evolution of Language*. Cambridge, Mass.: Harvard University Press.

Lineburg, Bruce. 1924. Communication by scent in the honeybee: A theory. *American Naturalist* 58: 530–537.

Lofting, Hugh. 1920. *The Story of Doctor Dolittle*. New York: Frederick A. Stokes.

Lofting, Hugh. 1922. *The Voyages of Doctor Dolittle*. New York: Frederick A. Stokes.

Lofting, Hugh. 1924. *Doctor Dolittle's Circus*. New York: Frederick A. Stokes.

Lofting, Hugh. 1925. *Doctor Dolittle's Zoo*. New York: Frederick A. Stokes.

Lofting, Hugh. 1926. *Doctor Dolittle's Caravan*. New York: Frederick A. Stokes.

Lofting, Hugh. 1927. *Doctor Dolittle's Garden*. New York: Frederick A. Stokes.

Lofting, Hugh. 1928. *Doctor Dolittle in the Moon*. Philadelphia: J. B. Lippincott.

Lofting, Hugh. 1933. *Doctor Dolittle's Return*. New York: Frederick A. Stokes.

Lofting, Hugh. 1948. *Doctor Dolittle and the Secret Lake*. Philadelphia: J. B. Lippincott.

Lubbock, John. 1874. *Ants, Bees and Wasps*. London: Kegan Paul, Trench.

Macedonia, Joseph M. 1993. The vocal repertoire of the ring-tailed lemur (*Lemur catta*). *Folia Primatologica* 61: 186–217.

Macedonia, Joseph M., and C. S. Evans. 1993. Variation among mammalian alarm call systems and the problem of meaning in animal signals. *Ethology* 93: 177–197.

Maeterlink, Maurice. 1901. *La vie des abeilles.* Paris: Charpentier.

Marler, Peter. 1999. On innateness: Are sparrow songs "learned" or "innate"? In Hauser and Konishi 1999.

Meier, Richard, D. Quinto, and K. Cormier, eds. 2002. *Modality and Structure in Signed and Spoken Languages.* Cambridge: Cambridge University Press.

Michelson, Axel. 1999. The dance language of honeybees: Recent findings and problems. *The Design of Animal Communication,* ed. M. D. Hauser and M. Konishi, pp. 111–131. Cambridge, Mass.: MIT Press.

Miles, H. Lyn White. 1990. The cognitive foundations for reference in a signing orangutan. *"Language" and Intelligence in Monkeys and Apes,* ed. S. T. Parker and K. R. Gibson, pp. 511–539. Cambridge: Cambridge University Press.

Morgan, Conwy Lloyd. 1894. *An Introduction to Comparative Psychology.* London: Walter Scott.

Morgan, Gary, Neil Smith, Ianthi Tsimpli, and Bencie Woll. 2002. Language against the odds: The learning of British Sign Language by a polyglot savant. *Journal of Linguistics* 38: 1–41.

Munn, Charles A. 1986. Birds that "cry wolf!" *Nature* 319: 143–145.

Nagel, Thomas. 1974. What is it like to be a bat? *Philosophical Review* 83: 435–450.

Nottebohm, Fernando. 1999. Anatomy and timing of vocal learning in birds. In Hauser and Konishi 1999.

O'Sullivan, Chris, and Carey Page Yeager. 1989. Communicative context and linguistic competence: The effects of social setting on a chimpanzee's conversational skill. In Gardner, Gardner, and van Cantfort 1989.

Owren, Michael J., and Drew Rendall. 2001. Sound on the rebound: Bringing form and function back to the forefront in understanding nonhuman primate vocal signaling. *Evolutionary Anthropology* 10: 58–71.

Pepperberg, Irene M. 2000. *The Alex Studies: Cognitive and Communicative Abilities of Grey Parrots.* Cambridge, Mass.: Harvard University Press.

Perlmutter, David M. 1991. The language of the deaf. *New York Review of Books* 38(6): 65–72.

Pesetsky, David. 1994. How to tell the apes. *New York Times Book Review,* December 25, p. 23.

References

Petitto, Laura Ann, and Paula F. Marentette. 1991. Babbling in the manual mode: Evidence for the ontogeny of language. *Science* 251: 1493–1496.

Pfungst, Oskar. 1911. *Clever Hans (the Horse of Mr. van Osten).* New York: Henry Holt.

Pinker, Steven. 1994. *The Language Instinct.* New York: William Morrow.

Poizner, Howard, Edward S. Klima, and Ursula Bellugi. 1987. *What the Hands Reveal about the Brain.* Cambridge, Mass.: MIT Press.

Povinelli, Daniel J. 2000. *Folk Physics for Apes: The Chimpanzee's Theory of How the World Works.* Oxford: Oxford University Press.

Premack, Ann James, and David Premack. 1972. Teaching language to an ape. *Scientific American* (October): 92–99.

Ristau, Carolyn A. 1991. Aspects of the cognitive ethology of an injury-feigning bird, the piping plover. *Cognitive Ethology: The Minds of Other Animals,* ed. C. A. Ristau, pp. 91–126. Hillsdale, N.J.: Lawrence Erlbaum Associates.

Rumbaugh, Duane. 1977. *Language Learning by a Chimpanzee: The Lana Project.* New York: Academic Press.

Russell, Bertrand. 1948. *Human Knowledge—Its Scope and Limits.* London: George Allen and Unwin.

Sacks, Oliver. 1989. *Seeing Voices.* Berkeley: University of California Press.

Sapir, Edward. 1921. *Language.* New York: Harcourt, Brace and World.

Savage-Rumbaugh, E. S. 1986. *Ape Language: From Conditioned Response to Symbol.* New York: Columbia University Press.

Savage-Rumbaugh, E. S., K. McDonald, R. A. Sevcick, W. D. Hopkins, and E. Rupert. 1986. Spontaneous symbol acquisition and communicative use by pygmy chimpanzees (*Pan paniscus*). *Journal of Experimental Psychology: General* 115: 211–235.

Savage-Rumbaugh, E. S., K. McDonald, R. A. Sevcick, W. D. Hopkins, and E. Rupert. 1993. Language comprehension in ape and child. *Monographs of the Society for Research in Child Development* 58: 1–221.

Savage-Rumbaugh, Sue, Stuart G. Shanker, and Talbot J. Taylor. 1998. *Apes, Language, and the Human Mind.* New York: Oxford University Press.

Seeley, Thomas D. 1995. *The Wisdom of the Hive.* Cambridge, Mass.: Harvard University Press.

Slobodchikoff, C. N. 2002. Cognition and communication in prairie dogs. *The Cognitive Animal,* ed. M. Bekoff, C. Allen, and G. M. Burghardt, pp. 257–264. Cambridge, Mass.: MIT Press.

References

Smith, Neil V., and I. M. Tsimpli. 1995. *The Mind of a Savant: Language-Learning and Modularity*. Oxford: Blackwell.

Spurzheim, Johann Gaspar. 1815. *The Physiognomical System of Drs. Gall and Spurzheim*. London: Baldwin, Craddock and Joy.

Stokoe, William. 1960. *Sign Language Structure: An Outline of the Visual Communication Systems of the American Deaf*. Silver Spring, Md.: Linstock Press.

Struhsaker, Thomas T. 1967. Auditory Communication among vervet monkeys (*Cercopithecus aethiops*). *Social Communication among Primates*, ed. S. A. Altmann, pp. 281–324. Chicago: University of Chicago Press.

Sturtevant, Edgar H. 1947. *An Introduction to Linguistic Science*. New Haven: Yale University Press.

Terrace, Herbert S. 1979. *Nim*. New York: Knopf.

Terrace, Herbert S., L. A. Petitto, R. J. Sanders, and T. G. Bever. 1979. Can an ape create a sentence? *Science* 206: 891–902.

Todt, Dietmar, and Henrike Hultsch. 1998. How songbirds deal with large amounts of serial information. *Cybernetics* 79: 487–500.

von Frisch, Karl. 1967. *The Dance Language and Orientation of Bees*. Cambridge, Mass.: Harvard University Press.

Wallman, Joel. 1992. *Aping Language*. Cambridge: Cambridge University Press.

Wenner, Adrian M. 2002. The elusive honey bee dance "language" hypothesis. *Journal of Insect Behavior* 15: 859–878.

Wenner, Adrian M., and Patrick H. Wells. 1990. *Anatomy of a Controversy: The Question of a "Language" among Bees*. New York: Columbia University Press.

Wilczynski, Walter, Harold H. Zakon, and Eliot A. Brenowitz. 1984. Acoustic communication in spring peepers. *Journal of Comparative Physiology A* 155: 577–584.

Williams, Heather, and Fernando Nottebohm. 1985. Auditory responses in avian vocal motor neurons: A motor theory for song perception in birds. *Science* 229: 279–282.

Zuberbühler, Klaus. 2002. A syntactic rule in forest monkey communication. *Animal Behaviour* 63: 293–299.

⇥ Credits ⇤

↠ Index ↞

Note: Italic page numbers refer to illustrations.

Primates (nonhuman) (continued)
See also Apes; Chimpanzees; Higher apes
Pro-forms, 207
Pronouns: reference for, 56–57, 211–12, 219; interpretation of, 60–61, 254; and American Sign Language, 249–50, 254, 258, 295
Prosimians, 179, 192
Protolanguage, 314–18, 323
Punjabi language, 123, 124

Quorum sensing, 16

Ramón y Cajal, Santiago, 158
Rapid fading, 23
Recursion: and human natural language, 8, 9, 11; and Kanzi, 12; and syntactic structure, 200, 295, 323; and phrase structure, 209–10; and expressive capacity, 230, 318; and chimpanzees, 278
Red-winged blackbirds (*Agelaius phoeniceus*), 130, 131
Reference: for pronouns, 56–57, 211–12, 219; and primate vocalizations, 167, 168; and alarm calls, 170, 191–92; and signed languages, 249; and apes' signing, 285; and artificial symbol systems, 287, 288, 290; and Alex's words, 303; and Kanzi, 320
Reflexiveness, and human natural language, 32–33
Regularities, 60, 126, 203, 242, 279, 297
Respiration, 25–26, 105, 107
Ristau, Carolyn, 40, 45–46, 48
Rules: and regularities of language, 60, 203; and phrase structure, 209–10, 213, 215; and analysis of auxiliaries, 213–14; and structural representations for sentences, 218
Rumbaugh, Duane, 287, 288
Russell, Bertrand, 14
Russian language, 123

Sainte-Laguë, André, 64
Sapir, Edward, 319

Sarah (chimpanzee), 267, 287, 288
Saussure, Ferdinand de, 241, 319
Savage-Rumbaugh, Sue, 35–36, 268, 287, 288–89, 293, 294, 296, 299, 320, 323
Screech calls, 146
Second-order intentional system, 190
Seeley, Thomas, 69–70, 75
Semantic signals, 26
Semantic soup theory, 278, 281, 296–97, 316
Semantic syntax, 29
Semantics: and human natural language, 26; and sentences, 210–11; and signed languages, 235; and syntactic structure, 295
Sensitive period: of birdsong learning, 148, 150, 151, 155, 156; and visual systems, 154; and human infant language development, 164, 317–18
Sentences: and syntax, 8, 201, 202, 211; sentence production, 49, 50–52, 162; ungrammatical sentences, 58–61, 201, 255–56; words and phrases of, 203–7; and phrasal types, 207–8; and phrase structure, 208–12, 213, 216, 218; and abstract syntactic structure, 213–22; and independence of syntax, 223–30; and signed languages, 255–56; Kanzi's understanding of, 292–93
Sequential elaboration, 261
Seyfarth, Robert, 32, 40, 171–72, 181, 187, 188, 190, 191–92
Shaping, 146, 154, 155
Sherman (chimpanzee), 268, 288, 289, 295, 299, 320
Signature Characteristic Strategy, 35
Signed English, 285, 286
Signed languages: and apes, 3, 11, 265, 266, 268; properties of, 11; structure of, 22–23, 35, 53, 235, 237, 254, 260, 265, 270; and vocal-auditory channel, 22; development of, 53–54; naturalness of, 54; and sound, 91; and human infant language development, 161–62, 258; and gestures, 233, 234; and ico-